How Culture Runs the Brain

Dialog-on-Freud Series

Series Editor: M. Andrew Holowchak

The Dialog-on-Freud series invites authors to explore the history and practice of analytic therapies through critical analysis of and expatiation on the seminal work of Freud. It seeks books that critically scrutinize the numerous facets of Freud's work over the course of his life, that investigate how or to what extent Freud's thinking causally gave rise to the various sorts of therapies that currently exist, and that examine the relevance of Freud's thinking today for those therapies.

Titles in the Series

Freud's Theory of Culture: Eros, Loss, and Politics, by Abraham Drassinower
The Unconscious without Freud, by Rosemarie Spooner Sand
How Talking Cures: Revealing Freud's Contributions to All Psychotherapies, by Lee Jaffe
Freud's Theory of Dreams: A Philosophico-Scientific Perspective, by Michael T. Michael
Nabokov's Mimicry of Freud: Art as Science, by Teckyoung Kwon
How Culture Runs the Brain: A Freudian View of Collective Syndromes, by Jay Evans Harris

How Culture Runs the Brain

A Freudian View of Collective Syndromes

by

Jay Evans Harris

LEXINGTON BOOKS

Lanham • Boulder • New York • London

Published by Lexington Books
An imprint of The Rowman & Littlefield Publishing Group, Inc.
4501 Forbes Boulevard, Suite 200, Lanham, Maryland 20706
www.rowman.com

Unit A, Whitacre Mews, 26-34 Stannary Street, London SE11 4AB

British Library Cataloguing in Publication Information Available

Library of Congress Cataloging-in-Publication Data Available
ISBN 978-1-4985-6245-4 (hardback : alk. paper)
ISBN 978-1-4985-6246-1 (ebook)

♾™ The paper used in this publication meets the minimum requirements of American National Standard for Information Sciences—Permanence of Paper for Printed Library Materials, ANSI/NISO Z39.48-1992.

Printed in the United States of America

Contents

Acknowledgments

My wife Maxine Susman edited the book thoroughly, critically, and faithfully through multiple drafts, influencing my understanding of American life, cultural trends, and women's experiences. She worked with me to articulate concepts in my inner mind that were difficult to formulate on the page. Aiming to reconstruct Freud's emotional discoveries and to harmonize his cultural influence with contemporary neuroscience, I knew Freud more from the inside out than the outside in, and I wanted to understand how he was impacted by Goethe's male version of culture and Darwin's understanding of innate bisexuality. Maxine helped me through this process and in distinguishing my own ideas from Freud's.

My longtime friend Kurt Langsten and I took parallel tracks in our professional lives, from training analyses in the 1960s to the serendipity of my taking over his ward at Metropolitan Hospital. We had lunch together weekdays for years, discussing Piaget's work as a neglected contribution to psychoanalysis, agreeing that Freud's work needed translation into neuropsychoanalytic science, and keeping in sight the work of neuroscientists like Gerald Edelman. Kurt has read my prepublished and published work with the tribal, sports-minded friendship that comes from sharing St. Paul, Minnesota as home.

In my frequent talks with Ty Heineken, who knows the world inside out from his lifelong travels and experiences as an anthropologist, we discussed developmental anthropology and finding identity within cultural mores, exploring the link between a subjective self and tribal morality. He helped me reframe my understanding of cultural identity in contemporary terms.

Peggy de Wolf and Jim Bogart lent ongoing friendship and support. They gave me exclusive use of their dining room table in Maine on summer visits, and we spent many evenings there and in New Jersey discussing the meaning of cultural life in light of current events and human evolutionary trends.

Rowman & Littlefield editors helped me shape this book. Amy King first invited me to contribute to a series on Freud's principles of cultural therapy. Molly White steered that draft toward a text of interdisciplinary scholarship. Kasey Beduhn oversaw its final production.

Introduction

This introduction is meant to demonstrate Freud's dynamic approaches to basic conflicts between culture and the individual. The entire book reconstructs Freud's lifelong devotion to understanding culture—and his ambivalence about taking on the task of elucidating cultural syndromes. His legacy joins his subjective experience to the objectivity of his reflection. A person's life work consists of projected fantasies of identity, meant to impact culture after death, because we feel kinship when others feel like our self and we do not want to believe our self can die. This book is an attempt to answer Freud's legacy wish: "… we may expect that one day someone will venture to embark upon a pathology of cultural communities" (Sigmund Freud, *Civilization and Its Discontents*, 1930, p. 144).

Here, we'll follow Freud's life crises and compensations into his cultural influences and cultural works. Since psychoanalysis was his subjective creativity objectified, Freud's ongoing memoir became the essence of his work. I will describe how his early childhood losses triggered narcissistic compensation, animating his theory that childhood development corresponds to archaic cultural development. Freud had the uncanny ability to logically objectify his subjective experience into a new science, a science of narcissistic reflection, building bridges to the external world of culture.

Let me be clear about the book's use of the term narcissism. There is no individuality without it. Narcissism refers to the degree of individuality, compared to the degree of cultural membership, composing feelings of identity. Immature narcissism refers to the intensity of wishing for one's own experience of pleasure, compared with concern for others. Leaders use narcissistic autonomy to make decisions for a group: while immature narcissists believe too many cooks spoil the broth, mature narcissists look for experts to help make decisions that are best for the whole group as well as for themselves.

Mature leaders have the capacity to reflect on experience in searching for better ways to use resources and to communicate their intentions.

The human brain evolved in a way that extends mammalian communication, including mammals' innate compassion for one another and distrust for strangers. Following Darwin, Freud surmised that people through millennia collectively built a cultural milieu to deal with survival and death. Faced with external danger or starvation, our archaic forebears turned to phantasies of imitating other people or animals for recourse. I'll use *phantasy* to represent collective myths—as Freud did, to distinguish it from individual *fantasy*. Collective rituals for dealing with severe stress converted religious phantasies into cultural beliefs.

Marcus Raichle, a pioneer in functional brain imaging, showed that the human brain's innate capacity for prediction evolved through the use of imagination when prediction failed. He discovered the brain's mode of thinking during failed prediction, calling it the *Default Mode Network* (DMN). Using imagination to enhance modes of survival when prediction fails, phantasy evolved in our brains as the source of making culture into virtual reality. I'll use the term virtual to refer to a phantasy shared by many people that then becomes cultural reality.

Freud's subjective affinity for narcissistic relationships with those who were self objects (i.e., experiencing another person as an extension of oneself) provided him with multiple perspectives on culture that he absorbed whenever he explored cultural topics, whether ancient or modern. We'll follow his reliance on Darwin and social Darwinism elaborated by other pioneer thinkers of his era, psychologist Wilhelm Wundt and anthropologist of religion William Robertson Smith. We'll also follow his identification with Goethe's approach to understanding the historical cultural development of the Holy Roman Empire as it influenced his Viennese-Germanic-Jewish modernism. We'll explore Freud's ambivalent relationship to America after World War I, through a close look at his collaboration with William Christian Bullitt on an analytic study of how Woodrow Wilson's religious delusions impacted European cultural regression.

Freud objectified his development of psychoanalysis through the brain science of his day. To give one example of my book's contemporary, neuropsychoanalytic view: The anterior cingulate cortex in our right brain activates in the same way when we are hurt physically or emotionally as when the same thing happens to our kin—a source of both phantasy and culture. But when prevailing culture becomes dissonant with individual identity, culturally induced repression sets in, setting up a dynamic of conflict between our wishes and the reality of our milieu. The thrust of this book is to examine how our own era has become trapped in a regressive process of breaking down the cultural hierarchy, in conflicts between global and tribal levels of culture; and to examine the interaction between terrorist groups and global economic leadership.

THE ILLUSION OF A FUTURE

In *Future of an Illusion*, Freud is critical of religion and sees it as sustaining an illusory belief in eternal life. Yet religions mediate conflict between individuals and culture in two ways: by soothing individual stress with rituals of social support, and by using institutional power to enhance members' political and economic resources. Ideally, the leaders of the religious hierarchy deal with the leaders of the cultural hierarchy, while a lower level of shamans and priests provide direct care to the members. As I see it, aligning my ideas with Freud's, all group members must demonstrate faith in a shared phantasy god, and sharing this symbiotic relationship to god extends the feeling of tribal kinship. Belief in souls, for example eternal cultural identity, is the fundamental illusion based on the brain's evolution of social phantasies. That belief blurs the distinction between culture and physical reality—perhaps the definition of a miracle. Belief in god's virtual power over all aggression is used to maintain the bureaucratic hierarchy of the religion. But aggression employed in sociopathic ways for personal or institutional gain by immature narcissistic leaders inside or outside of religion sometimes leads to crusades or jihads, resulting in cultural regression.

Immature narcissism and oligarchy in many nations lead to global regression, and inevitably to epidemics of cultural syndromes enacting collective phantasies of *restitution* and *retribution*. This book examines those syndromes by applying Freud's understanding of three kinds of thinking necessary for individual and group survival: practical, fantasy, and emergency thinking. To my understanding, each mode entails a separate brain state responding to metabolic needs and external safety needs that evolved as changing milieus required new social cooperation and restraint. Freud added a key principle to Darwin's theory of evolving mammalian communication: human childhood stages were modified in response to long eras of cultural change. From Freud's perspective, culture evolves to control (bind) an individual's raw, instinctual libidinal and aggressive instincts.

In Freud's view, culture evolved from primal hordes to extended families, tribes, kingdoms, nations, and global organization in hierarchical stages, with each new stage requiring increased communication with strangers. In his model of the instincts, the brain seeks appetitive satisfaction among familiars but prepares for aggression toward strangers. But aggression is bound by cultural needs, since survival depends on learning the techniques of strangers and bonding with them in higher-level cultural structures. Throughout recorded history, if oligarchs are too selfish, fulfilling their wishes at the expense of existing subcultures, individual aggression in those subcultures unbinds and cascades into cultural regression.

HOW FREUD'S LIFE EXPERIENCE DETERMINED HIS CULTURAL THEORIES

Understanding the effect of immature, narcissistic leaders on world culture benefits from studying how Freud made his discoveries about narcissism and forms of identification. His discoveries arose from the development of his maturing narcissistic talents and from his reflection on what he learned through a lifetime of self-analysis. Early childhood losses led him along with his nuclear family to a whole new environment, triggering narcissistic compensations that energized his cultural theories. I will summarize what he revealed in self-analysis, which I stitched together in a biography of Freud which shows that his early childhood loss led to a life of restitution: attempts to make culture whole again (Harris and Harris, 1984).

While Sigmund was still an infant, his mother Martha was mourning the death of her brother Julius. Naming her next child Julius, she became depressed when he died at seven months, when Sigismund was almost two. Moreover, her husband Jacob's failing textile business undermined the businesses of his two sons from an earlier marriage. His older son Emanuel had two children, a girl Sigismund's age and a boy a year older. He, his wife, his children, and his 20-year-old brother Philip migrated away from Jacob's family in Freiberg, Czechoslovakia to England when Sigismund was three.

To make matters worse, Monika, the Catholic woman who helped care for all the youngsters, had been found guilty of stealing Sigismund's birth-gift money. Sigismund was deeply attached to her, especially in his mother's absence. Monika had been taking Sigismund to church and telling him about the hell beneath the ground—where baby Julius had been buried. The mayor of Freiberg, who was also the Catholic family doctor—a man with only one eye—attended the sick baby Julius and presided over his burial. Later, he stitched up Sigismund's chin when he fell searching a cupboard for his pregnant mother. Freud's self-analysis and many dreams pointed to this imagery as the foundation of his early childhood sense of loss, and split between a prevailing Christian culture and a family-oriented Jewish culture. Leaving Freiberg on a train to Leipzig with his mother, he saw gas jets on fire from under the ground—proving Monika's lessons true. Sigismund must have felt that non-Jews controlled life, death, and money in a way that divided his family.

Failing to revive his business in Leipzig after a year, Jacob and Martha moved their expanding nuclear family to Vienna where her family helped financially. Sigismund could not encompass his early childhood losses; his individuation split between a sense of identity before Vienna, and a later one. The mature analyst Freud knew that children cannot grieve, because they do not have a context for what they have lost; instead they experience narcissistic compensation for what seems missing from self-identity, for example, a tendency throughout life to seek self-extending cultural structures. From that perspective, his losses led to his determination to create a progressive culture through his analytic work. Yet preoedipal survivor guilt led to so much compassion for others who suffer loss and trauma, that he manifested both narcissistic depression and oedipal survivor pride.

WHAT LEADS CULTURES TO RELEASE AGGRESSION?

Freud explored this question his whole life long. He wanted to make psychoanalysis a source of understanding in reconciliation with the prevailing culture. His famous dream image in *The Interpretation of Dreams*: "You are requested to close one eye" after his father Jacob's death, showed his attempt to close the split in his identity (Freud, 1900, p. 317). He wondered how the wish for others' death arose in the archaic cultural past. Writing out his self-analysis in letters to Wilhelm Fliess, his muse for his book on dreams and his self-object pseudo-analyst, he split his death wishes conceptually. He mixed preoedipal conflict based on primal ambivalence toward Julius who took the breast from him, with his discovery of the oedipal wish to kill his father (or the one-eyed-doctor). His fantasy of making psychoanalysis into a cultural force used muses like Fliess and Jung, revenants of Julius—to show his views as more mature than theirs.

Freud's split between early and later childhood led him to confuse the timing of essential development: Is oedipal guilt innate in infancy or only in oedipal stages from three to five years old? Even now, psychoanalysis splits narratives of dynamic development between object relations analysts who follow Melanie Klein in seeing oedipal conflict inherent in infancy versus Anna Freud's followers, who use ego psychologists' child observations to buttress their conclusions of oedipal periods from three to five. Showing how stages build earlier structures of identity organization into later ones, Freud thought many dreams retained a navel: innate archaic dynamics.

His developmental context of brain, mind, and culture gave equal place to libido and aggression during all stages. But he found clinically that derivative oedipal fantasies were intrinsic to an individual's cultural development. Six- to nine-year-old latency-age family romance fantasies of noble alternative parents were often strong in the daydreams and ambitions of would-be heroes, culture builders, sociopaths, and impostors, as well as paranoids and psychotics. Since Freud felt family romance identification with Christian forebears—especially Goethe, I will show what he learned from him about how to influence culture, and how to write about it. Indeed, Goethe brought world literature, including Gilgamesh and the Bible into the scope of Freud's cultural modernism.

CULTURAL SYNDROMES: RESTITUTION AND RETRIBUTION

As I see it, *restitution* and *retribution* are cultural syndromes that result from unbearable loss and trauma, respectively—often coexisting. Freud surmised that regressive fantasies during the "universal" *endogenous* trauma of passage to new stages may trigger neurosis or character disorder—behavior different from the cultural norm. But *exogenous* trauma or loss during any stage may trigger regressive denial of reality, creativity, or sociopathy. In trying to unify his anxiety-theory of how syndromes form, Freud may have glossed over the effect of cultural syndromes on individual maladaptive thinking.

Cultural disasters, like World War II concentration camps, caused unbearable loss that spread to future generations. Primo Levi describes character traits that allowed some individuals to survive: (1) a talent for language, (2) an affinity for bonding with strangers, (3) a sociopathic capacity to ferret out opportunity, and (4) a reflective ability to bear witness (Levi, *The Reawakening*, 1995). Survivors were the leaders in their camp. These extreme adaptive qualities merging mind, brain, and culture during prolonged emergency differ from both PTSD and practical thinking. These qualities may develop heroically or pathologically in talented or vulnerable members of syndromatic cultures. Levi observed that many survivors could never recover their familiar, settled identity when they rejoined kin who had migrated or entered safer refugee camps. Those who emigrated to the United States of America moved on to affiliate with patriotism when they were able to join their new social and economic world, and were less likely to promote a cultural syndrome of refinding their past through restitution fantasies.

In present-day economic deprivation, mass migration, refugee camps, and chronic warfare, many cultures develop multigenerational cultural syndromes. Collective longing for cultural integrity conflicts with requirements for acculturation to new language, place, bureaucracy, opportunity, and morality, as well as with new dominant religions, tribes, and ethnicities. Perhaps every culture has had its own syndromes of failed acculturation giving rise to myths. From that perspective, myths condense founders' and forebears' loss and trauma as truths and fictions to build progress on. For instance, even the godless in the French Revolution hid their sociopathic sins against one another and glorified their collective survival by building a pantheon mythology of their greats.

SUBCULTURE PROBLEMS OF BIAS

When subcultures in a national hierarchy represent the classes at the economic bottom, bias intervenes as a rationale for maintaining forms of slavery, caste, or class. This book's neuropsychoanalytic perspective shows innate bias arising in the infant brain's love that binds kinship and aggression toward strangers. Thus, the orbital frontal cortex, which remains the first take on response to others before we know it in adult life—bias, is reconditioned in young childhood and in all the stages after. Thus, conditioned early childhood bias can be overcome later through cultural morality that applies regulated law to give all subculture members equal opportunity to survive.

It seems that subcultures, including those with their own syndromes, have different norms for their stages of bias, depending on their collective phantasies. The rationalized role of bias in unbinding aggression between diverse subcultures due to failure of economic equality from the top-down does not figure into the statistics that measure economic global or national well-being. Wars are not useful for progress as they were thought to be among colonial powers during Freud's pre-World War I life, any more than democracy can be exported now if grassroots are not deep enough to hold.

To explore cultural syndromes let us consider rationales for bias that maintain political and economic power. The U.S. Supreme Court decision that corporate groups have the same legal rights as individuals can unleash aggression against individual rights. Influences within the law are presented by corporate lobbyists as moral guides that provide benefits for interest groups that want to maintain survival advantages that stem from working within national, state, or local bureaucracies. Libertarians who want Free State's rights to maintain their individual lives as they see fit do not acknowledge how much their freedom benefits from national contributions to their welfare from technology protected under national law.

Later in the book, exploring how narcissistic individuals influence culture in their own or their buzz group's interests, we'll consider how contemporary lobbyists, media, journalists, bloggers, advertisers, and others plant influence as cultural rationalizations that have a seeming moral thrust. Mature narcissists such as start-up, Silicon Valley, types of entrepreneurs use joint enterprise to influence cultural economic trends. They often claim that they help culture in the way that Freud thought was progress. Internet-based social influence on what products to buy may determine a subculture's reputation. Older and younger people are also targeted players in the interest group sweepstakes.

Inevitably, economically underprivileged subcultures have more members ready to enter cultural syndromes. Consider Black Americans growing up in a chronic multigenerational subculture of hopelessness about the future. A lack of equal opportunity to enter the higher economic echelons led toward membership in a syndromatic subculture seeking restitution. Given their long-term, historical, multigenerational victimization as slaves lacking full citizenship subject to the economic wishes of a hierarchical culture that excludes them from full membership, only a few in each generation can find a way to join the economic hierarchy as full-fledged members of the U.S. culture.

Those circumstances led both blacks and American Indians to be members of subcultures denied access to equal freedoms. Concentrating here on the black male subculture in America, lacking freedom to develop a wide array of their talents because of bias, individuals may use drugs to feel identity pleasure or join gangs that urge them to take what should belong to them. In families broken by poverty, illness, violence, or lack of education, alternate identity is offered through religion, sports, music, or gangs. Black males entering the prevailing American economic hierarchy, feel conflict in distancing from their syndromatic culture. Consider the role of black cops. When I was a medical student in Baltimore in 1960, I compared black and white kids using black and white puppets, getting the children to talk to the puppets about social situations. The younger black children identified with the TV media's prevailing white culture, until older peers in the neighborhood taught them to identify with their own subculture.

RETRIBUTION: SYNDROMES OF IDENTIFICATION WITH THE AGGRESSOR

Retribution often leads to trauma and reliving identification with the aggressor can lead to terrorism. On a locked ward at Metropolitan Hospital in New York in the 1980s, where I was Ward Chief, we had a patient who attacked and threatened other patients and staff. I asked him why he acted as if he wanted everyone to be afraid of him. He said that for years when he was a kid, his uncle—guardian—beat him and his younger sister when they were disobedient, locking them in the basement for days without food or drink. He felt vengeful toward anyone who reminded him of that—especially in a locked ward.

Another patient was admitted to the ward with severe anxiety. When I approached his bedside with a group of students to meet him on the morning after his night admission, he suddenly jumped from the bed, crouching on the floor in front of us—ready to strike. We learned that 15 years before, as a 20-year-old soldier in charge of 19 young men on a boat, he was so afraid of losing control that he mowed them down. He developed PTSD with panic and repetition-compulsion. You may recall the famous photo of Patty Hearst, kidnapped by the Symbionese Liberation Front, holding a gun while they robbed a bank. The leader of the symbiotes in that cult of revolution got her to identify with their aggressive purpose as a member. She later married her body guard.

THE BRAIN IN EMERGENCY MODES

The brain thinks practically whenever it can. Its left brain (LB) initiates appetites for satisfying metabolic and sexual needs, while the right brain (RB) responds to reality conditions. Psychologically, this division provides LB subjective and RB objective sides of self. The LB mode triggers the formation of narcissistic identity—individuation; while the RB mode objectifies social identity. The syndromatic brain shows unusual LB and RB activity in loss and trauma. In depression, the LB shows less pleasure activation and in trauma the RB presents less social and cultural identity activation (Northoff, 2011). If the LB promotes pleasure drives toward productive goals and appetitive satisfaction, and the RB promotes reality perception, Freud's pleasure principle and reality principle dominate separate hemispheres.

This book discusses two forms of cultural syndromes, LB *loss and restitution*, and RB *trauma and retribution*. In practical thinking, the RB maintains sympathetic nervous system vigilance, evoking *frustration* when reality presents failed salience in expected satisfiers or *signal anxiety* over possible danger. It triggers the brain's *default domain*, when frustration requires reassessment in the thinking mode of fantasy and self-meditation (Lanteaume, 2007). Emergency thinking begins when unbearable stress and fear cause the right amygdala to block traumatic experience from entering hippocampal long-term-memory and high cortisone blocks the emotional responses that support self-identity. In prolonged acute stress the right amygdala triggers instinctual aggression: fight, flight, or paralysis.

During such emergencies, changes in the hormonal and phenotypic milieu block the reception of experience, bringing perplexity: self-integration is suspended. Prolonged emergency may result in a lasting PTSD syndrome. You don't feel like the same person. After a month without treatment or medication, PTSD may become entrenched. Keeping veterans waiting is a travesty. Feeling a lost capacity to love or be loved many veterans suicide. And on a collective basis, PTSD can spread culturally like an epidemic.

THE CULTURAL EVOLUTION OF THE HUMAN ANIMAL

Cultural regression reverses the stages in the development of global culture. As the global economic equilibrium that supports survival in most nations succumbs to the inequity that provides a disproportionate share to the very rich, the competition among nations and among economic subcultures increases. People turn to their religious sects and their tribes to find collective help in reversing the threats to their well-being and survival. Protests and demonstrations in public places test the coherence of government, the cooperation among divergent factions, and the collective will of the people.

Inevitably, the more sociopathic and revolutionary people join others who are already vulnerable to acting out phantasies because of living in syndromatic subcultures of their own. Just as wishes trigger fantasies and REM dreams, collective phantasies embedded in family, tribal, and religious mythology can be enacted. Such phantasies emerge after genocide, starvation, disease, forced migration, and life in refugee camps. Losing kin, ethnic and religious community, and homes and customs—everything familiar—induces cultural syndromes among survivors. In *The Third Chimpanzee*, Jared Diamond, a popular anthropologist, documented how frequently the world bears witness to mass genocidal murders—in every other year after World War II (Diamond, 1993, p. 284). Human cultural evolution has been a blessing and a curse, that is, both progressive and regressive.

The human milieu evolved from mammalian origins. There is evidence that the human medial brain evolved differently from the brains of other mammals in processing group relationships among individuals, while the lateral brain evolved complex cognition. Freud viewed individual conflicts as caused by developmentally layered, culturally repressed, fantasy wishes still seeking consummation. Clinically he thought individual syndromes developed when repression was too harsh. In trying to understand the sources of human mental pathology, he concluded that cultural identity evolved from mammalian brains and continued to evolve in stages of cultural evolution. The LB evolved language capacity for behavioral sequences of action, while the RB evolved affects to control behavior. As an individual moves from infancy to young childhood, Freud's pleasure principle gives way to reality.

Many neuroscientists believe the brain to be an intricate mass of molecules and ions, responding to external energy. That view avoids the medial brain's sensitivity to ongoing shifts in relational forms of communication. Many neurons have multimodal receptors to a variety of sensory and emotional inputs. Those who believe the brain is simply physical structure do not account for its adaptability to cultural input. Thought and culture evolved in cultural stages of imagining a cultural world populated by gods and the dead. Without imagining virtual beings, the brain could not predict how best to survive.

HOW CULTURAL EVOLUTION CORRESPONDS TO THE BRAIN'S PSYCHOLOGICAL DEVELOPMENT

Freud wrote that spontaneous group behavior corresponds to how primal hordes follow their leader, much in the way that herd animals follow theirs, or a person in hypnosis allows a leader to preempt their individual will. This, you could say, corresponds to the cult in culture. Freud imagined archaic hordes following a leader like infant symbiotes. Members of fundamentalist cultures including Christians, Muslims, or Jews, replicate the infant helplessness of symbiotes. I will present two examples of contemporary tribal values, a woman and a man who show the conflict in the contemporary world between tribal and global values.

My patient was a highly Orthodox, 45-year-old Jewish woman living in Brooklyn who suffered hypochondria, anxiety, and an inability to make decisions. The second youngest of 14 siblings, she was bright and precocious, but her father blocked her wish to continue education beyond high school and arranged a marriage for her with a sadistic man. After she gave birth to her third child, and had a prolonged postpartum depression her husband attempted to kidnap the children and take them to Israel. Her father put a stop to it and she was given a Jewish divorce.

Yet her ongoing, moody anguish was considered a bipolar sign. Her father insisted she take lithium to calm down. Although lithium was stopped 20 years before her treatment with me, her physical complaints reflected kidney problems. In her therapy, she knew she needed medical attention but refused to see a medical doctor until a highly placed rabbi would recommend she do it. Since her father's death she would agonize over decisions for months on end until she could justify seeking rabbinical backing. She lived on welfare despite having a wealthy family who would give her money to see rabbis or for religious occasions.

Modern access to the world outside of her faith was forbidden. It was out of the question to have a TV or a computer, but she had a phone, which she used to get recorded daily religious messages. She believed in the future coming of the Messiah and life after death where she would join her dead parents. Those beliefs animated her compulsive prayers, recited for each segment of every day. On the Day of Atonement she would beg forgiveness from everyone she had felt hurt by during the past year—because she may have shown anger.

Discussions about seeing a male doctor—it had to be male—would lead to her avoidant denial of medical help. She insisted I help her resolve her anxious indecision with methods of relaxation techniques, special diet, and prayer. Finally, she admitted guilt for having felt attracted to a man whom she heard singing Jewish songs at a wedding. She believed those feelings were the source of her anguish and poor health. After consultation with a high rabbi that took months to plan, she saw an orthodox kidney specialist who told her she needed a kidney transplant.

While agonizing over whether she could use her 22-year-old son as a kidney donor, rather than that of an unknown donor, an incident happened that led to the end of her therapy. Having a rented office, I found my door locked before her appointment; and I did not have a key for the upper lock. There was no office to see her. So, being close to Prospect Park, we had the session on a park bench. It seemed to go well as she saw her way clear to accepting any kidney donor. At that point her orthodox kidney specialist insisted she see his nutritionist and his social worker for therapy. A few sessions later, she said the session in the park had been against her religion. That was our last session. Her kidney specialist had become her leader.

My patient's lack of individuation was heavily influenced by the suppression of sexuality and aggression demanded by her Orthodox community through her father's power over the family and his high status in the Orthodox community. Just as devout Muslim women cover their hair, my patient wore a wig. For Orthodox Jews, a woman's hair represents her sexuality. She believed that the enjoyment of feelings of love, or even the pleasure of music, was a step in the direction of sin; enough to lose the right of eternal life. Women's rights to their own sexuality threaten the male cultural dominance that is most compelling in fundamentalist cultures.

Very orthodox, ultra-conservative, fundamentalist religions and religious sects tend to maintain archaic forms of membership in a culture dominated by a male leader who deals with group survival by regulating group behavior. The leader reflects on his group's milieu, using cognition to recognize how they must behave to remain survival safe. Over evolutionary time some tribal leaders tend to develop or overdevelop their own narcissistic exceptionalism, while members feel symbiotic group identity—like cardiac muscles that function together; yet separate in their roles. Group affect, as Freud showed, spreads cohesion among members that expresses the identity of each as a cultural unit.

In the contemporary world, leaders of fundamentalist cultures have maintained their economic clout. In the Muslim, Christian, Jewish, and Hindu worlds, for instance, fundamentalist leaders can become oligarchs at the same time that they support religious tribalism. Some degree of sociopathy or entrepreneurship can be found at all levels of the cultural hierarchy, especially at the top and bottom of national economic and political systems. After talking to two of my patient's brothers during her crises, it was clear that following their father, the men in her family participated in the global world's economic life, as well as maintaining their orthodoxy.

THE STRUCTURE OF FUNDAMENTALISM

Freud believed that culture controlled the sexuality of its members in order to enforce the power of its leaders—a theme this book explores. Although women are highly controlled in fundamentalist cultures, men too are enslaved to the feeling of being taken care of by the community—now, or in an afterlife. My patient's absorption in her community is akin to the 2015 memoir by Shulem Deen, *All Who Go Do Not Return*. Deen gradually deviated from the fundamentalist life of a Jewish Skiver Hasid. For him too, individual reflection was banned because all he was allowed to know had to be based on Talmudic words of the founders of his sect about every aspect of life. Studying the Talmud over a lifetime and following its edicts is like memorizing the Quran, or obeying a Latin Liturgy.

This book examines how imitation as obedience fulfills archaic membership. When Deen was 18 and assigned a wife, he had only 7 minutes to meet her before the wedding six months later. Knowing nothing about relating to a wife or intercourse, he learned from male community leaders that a wife is not meant to be a friend, but a community help-mate to ensure that everything is done properly. His rabbi explained sexual activity to him—what must be done physically, how it must occur every Shabbat night except when his wife had her period—when she was untouchable. Indeed, all life functions, including the preparation and eating of food was defined with compulsive exactness. To veer from kosher was as damaging as a tribal member breaking a taboo. Curiosity about the world outside his community, even about other Orthodox sects was sinful and liable to attacks—like one he had seen enacted by his whole community as their lead rabbi posed danger when outsiders drove into their village looking for fun.

The sociopathy of the leaders of Shulem's cult showed up in their scams of money collected from government agencies for welfare, food stamps, and education—mainly Talmudic. Male members were in a double bind because they could not afford to support a household with as many children as possible. For weddings, circumcisions, Bar Mitzvahs, holiday rituals, and special advice, leaders expected significant gifts. To get enough money to support a growing family, Deen had to find a profession beyond teaching Talmud, since rapping recalcitrant boys on the knuckles with the required sadistic vengeance ruined his self-respect—that degree of individuation led him toward the outside world. His determination to develop a capacity for work in the outside world, ineluctably led to loss of his wife and children, who remained enmeshed in their sect. Excommunicated, he received the same aggression as any outsider.

Freud viewed the regulation of love and aggression as ways that culture controls individuals. Community obedience blocks individuation. Meaningful identity seems to be the fulcrum between the individual and cultural morality. Freud realized that cultural regulation of individuation and sexuality opposed the maturation of individual talents. In modern culture feminists berate their lack of cultural status. Yet, in *On Narcissism* (1914), Freud presented female narcissism as self-defined in terms of their appeal to men as sexual and reproductive objects. We'll see how Lou Andreas Salomé set the stage for changes in Freud's understanding of female development and sexual identity, and how Anna Freud and her partner Dorothy Burlingham showed Freud how women—not only men—work together to enhance cultural and psychoanalytic development.

HOW MY PROFESSIONAL EXPERIENCE INTRODUCED ME TO CULTURE

When I was a Fellow at the Albert Einstein College Hospital in the Bronx in the late 1960s, our new psychiatry ward was modeled on the famous Chestnut Lodge, where inpatients could be treated for years if necessary. New staff and patients worked together toward creating their own therapeutic subculture. Staff met weekly to discuss the ongoing therapeutic milieu: dynamic psychotherapy four times weekly, group therapy, and art therapy—including the making of weekly group murals, run by an art therapist, Cliff Joseph, who noted each patient's participation in choosing a collective mural theme, and then spontaneously adding to the group mural. This work led to *Murals of the Mind* (Harris and Joseph, 1973).

What a revelation of creative process! A repeating sequence of murals created over six- to seven-week periods tracked the cultural stage of group development. Each stage began with a theme of conflict; the following week's mural showed the conflict as a problem causing group anxiety. The next week, an underwater or underground scene manifested unconscious fantasy. Next, resolution of the conflict was portrayed through humor and relief; and in the last mural of the sequence, the group enjoyed its coherence: problem solved.

The formation of group ego structure in stages corresponds to the sequence of developmental stages in childhood. Symbiotic group formation was followed consecutively by a stage of control issues, then competition for leadership, and finally reflective group history and myth-making to recapitulate what the group had gone through. Minutes of the weekly staff meetings during that period showed the same stages in staff development. So it seems that, like individuals, groups develop from symbiosis to control to competition to reflection.

Using the same model in this book, I will examine how culture progresses and regresses through these developmental stages. *How Culture Runs the*

Brain explores the regressive effect of immature narcissistic leaders, compared with the progressive effects of mature narcissistic leaders, on the cultural organization of the global world. It applies Freudian principles to the mechanisms of progressive and regressive stages of individual and cultural development as they are mediated by individual and collective phantasies. In members of fundamentalist cults, the symbiotic fixation to infancy stunts individuation. Instead of making a passage from infancy to young childhood that induces ownership of one's own identity, the cult still owns the member. The idea of "Mine" develops in young childhood autonomy. Without sufficient entitlement, a person may remain enslaved to leaders due to anxiety about abandonment similar to what is felt in young childhood. Immature narcissists who become world leaders maintain their infantile grandiosity and the "mineness" of young childhood.

Every theory of mind acts as a bridge between one's biological self and others' biological selves via the linguistic bond of familiarity or estrangement from one another. The medial networks of the brain sustain human communication by using language to reflect culture within the mind. Given that self-observation reflects one's social and cultural standing, individual identity is as much determined by culture as by biological signals and instincts. Male narcissism, which historically has been used as the legal arbiter of cultural and economic power, will be explored as a major cause of unbinding aggression in the formation of cultural syndromes.

This is a view of cultural evolution arising within the medial networks of the brain, in minds that imagine reality—including the reality of other minds. During developmental stages, identification with others and fantasies about them produce so many imagined ways of being that culture becomes a virtual catch-all. For instance, Freud's discovery of the male fetish as an imaginary female penis that relieves castration anxiety in some men, projects collectively into culture as a concept of virtual value. Freud learned a lot from Goethe about the role of the fetish in culture. Goethe's *Faust*, often quoted by Freud, dramatizes how male narcissism projects sexual identity bias onto females. That is one example of the virtual world entering into the deepest sources of instinctual pleasure.

Freud discovered that latency is the time when children set cultural identity into their character structure. He showed how family romance fantasy (the idea of having alternative parents) brings preoedipal and oedipal fantasies into a new structure of cultural heroism and mythology. Following after Freud, my book explores how, in our contemporary world, latency fantasies of ambitious heroism thrive in syndromatic cultures and how children in these cultures absorb grandiose and revolutionary phantasies of restoring lost cultural identity.

Chapter 1

Contemporary Cultural Syndromes

Freud blamed the false beliefs of grandiose leaders for ruining cultural progress. In particular he blamed President Wilson and the Wilhelm II for pursuing grandiose plans to run the world as they wished under cover of religious authority. He says believing god was on their side in World War I, the Allies and the Central Powers both used religion to hide their grandiosity. In the introduction to a book Freud wrote with William Christian Bullitt, he blamed Wilson whom he quotes:

> *God ordained that I should be the next President of the United States* ... a man ... capable of taking the illusions of religion so literally ... is unfitted for relations with the ordinary children of men. ... the hostile camp ... also sheltered a chosen darling of Providence: the German Kaiser. ... (Introduction by Sigmund Freud to *Thomas Woodrow Wilson: A Psychological Study*, 1966, p. xi)

He blamed America for weakening the central European economy by failing to deter the Allies from requiring impossible German reparations after World War I. Chapter 1 follows the collective narcissism of leadership into its contemporary failure. When global leaders pursue their own advantage, tribal and religious subcultures regress into enacting their own archaic phantasies. There is natural conflict between hierarchical leadership and the experience of those who are led.

Twenty-first-century America still claims god as its source of economic greatness. Its financial leaders used the dollar, the global standard of value, to sell derivatives, vastly exaggerating the value of its real estate. Profits went to the world's large banks, bankers, and investors. The resulting Great Recession undermined the world's poor and middle classes, triggering a global regression cascading into an epidemic of unbound aggression. Many of my Freudian and empirical ideas about cultural syndromes have intensified in the years since I wrote *Minding the Social Brain* (Harris, 2013). Freud's cultural works can help us understand how undermining the global economy leads to cultural syndromes. We'll assess cultural challenges that narcissistic oligarchs generate with their money and influence and how those challenges contribute to terrorism.

Given wars, migrations, and collective PTSD (post-traumatic stress disorder) going back generations through the past century, we can discern two kinds of prevailing cultural syndromes that correspond to Freud's understanding of the effects of loss and trauma: collective wishes for *restitution* and *retribution,* respectively. When the global world enters economic recession, affected nations regress along their fault lines, unbinding aggression among sectors of their culture that can lead to civil war—intensified by other nations. Syria's civil war shows how contemporary religious and tribal conflicts unbind aggression. Trauma, genocide, and mass migration trigger cultural syndromes that disturb the multicultural equilibrium.

For another example, consider Palestine's attempt to become its own nation by turning

Hamas's choice of leaders toward state-building, and away from using aggression to support Jihad. The British journalist, Sara Helm writes that incursions by Isis threaten this state-building. She describes Palestinian vulnerability to Isis because their children were affected by the trauma of war, occupation, and encirclement by Israel, from one generation to another (Helm, January 14, 2016, p. 20).

Children, who see parents and political leaders unable to save their home from destruction, turn to enacting imbedded revolutionary phantasies. Chronic cultural dilemmas trigger severe stress and regression to early developed childhood thinking in vulnerable individuals. Let's consider global cultural pathology by contrasting immature with mature narcissistic leaders.

FREUD'S VIEW OF NARCISSISTIC INDIVIDUATION

For Freud, mature narcissism meant understanding that sharing satisfaction enhances both individuals and culture. Mature narcissists are proud to influence overall cultural progress. Their modus vivendi uses multiple reflections on their life experience to widen their scope of knowledge and they share that knowledge with their cohort of other mature narcissists. Freud's reflective understanding of his own mental experiences as it changed over time provided the apercus he needed to create a movement by using his talent as a writer. He impacted culture by founding journals to bring other experts into collaboration.

In the modernism of the late nineteenth century, journalists gathered information from economic leaders; understanding social trends became recognized as the means of predicting success. Those able to predict changes in technology and social policy had the competitive advantage in gathering resources. As we'll see in later chapters, Karl Marx and Theodore Herzl also used journals to impact culture. In Freudian terms, mature narcissists enjoy their oedipal urges by sublimating them, not acting them out as immature narcissists do.

Yet, mature narcissists, like members of most subcultures have their blind spots about culture. In Freud's time, genocide and forced migrations were rationalized by the politics of colonialism that relegated tribes, ethnic groups, and their religions to inferior cultural status when they were colonized. As WW1 approached, various colonized groups were determined to regain cultural status by forming their own national entities. Freud believed that nations of varied subcultures were more progressive than uniform tribal or religious cultures. He recognized that when the Allies carved up Europe and the Middle East, they disregarded how ethnicities worked together; thus selectively weakening some tribes and strengthening others. And by exacting impossible reparations on Germany, they set the stage for World War II.

The inherent divide between tribal cultures and global culture corresponds to the divide between symbiosis and diplomacy. Love for one's kin is so much more satisfying and familiar than love for one's country or for the whole world. Hate is more compelling than diplomacy in dealing with strangers. Progress in hierarchical cultures formed of mixed subcultures requires individuals to reflect and learn from others to form cognitive truths. Collective mature narcissistic reflection provides the bridge between tribalism and globalism. But just as too many cooks spoil the broth, too many individual leaders give rise to some immature narcissists and sociopaths who contrive to lead their tribes to states or nations for their own benefit.

Looking at the development of mature narcissists in our own day, the politics of influence have amplified their reach. Having replaced the more impartial past journalists, speech writers, researchers, and representatives of tax-free, legally nonpolitical agencies comprise a vast contingent of lobbyists who virtually run government policy: "Members of congress are frequently unaware of financial ties between industries and those with think tank titles acting as witnesses at hearings. Indeed, the hearings determine economically-related laws rigged by corporate influence" (Lipton, Confessore, Williams, *The New York Times*, August 8, 2016, p. 1).

LOSING CULTURAL REALITY: AMERICAN CULTURAL REGRESSION

Twenty-first-century American politics often invokes democracy and religion as moral proof of leadership that undermines national law. In the prime example, freeing Iraq from its dictator, as if that could trigger democracy, rationalized policies aimed toward leaders' economic gain. Using sanctimonious words to hide selfish gain is symptomatic of cultural regression. Hidden policies for government action, such as the CIA falsifying reports on its incursions; spying on the public and global world leaders; water-boarding captives as if they were legal war criminals, all exemplify the U.S. legal hypocrisy that gave impetus to Al-Qaeda and Isis. Projecting aggression as if comes from an enemy is a form of moral intimidation that invites attack.

As I see it in 2017, the United States has regressed politically to the point of losing a collective sense of its national cultural identity. By 2012 the Grover Norquist Taxpayer Protection Pledge of the Republican Party was not a pledge of allegiance to the nation, but a pledge of aggression against Democrats who would otherwise share too many national resources with poorer subcultures. It was used to block most of the legislation proposed by Obama to help America emerge from the Great Recession. The pledge felt necessary, because it insured reelection funding from the republican elite.

Republican rationalizations about budget constraints directed against the poor were treated by media as competition for leadership. Political compromise was replaced by bias—a form of multicultural lying. Trying to tell the truth to the American public, Obama was isolated and held responsible for failed policy by Republican stonewalling: They would not cooperate on anything, while the democrats seemed content to withdraw from public life to keep the peace or to join the ranks of the mature narcissists. Infected by a top-down cultural epidemic of projected aggression, many people dehumanized Obama, blaming him for the increase in terrorism. The widespread availability of weapons was symptomatic of unbinding aggression, as if war weapons might be needed if the national government tried to repeal personal liberty. Norquist had also represented the NRA gun lobby, which donated large sums to reelection of gun-minded Republicans. The NRA lobbied for many state legal rules to block or gerrymander lower-class democratic voters.

To the extent that Republicans projected their aggression onto President Obama, claiming that he was the reason no legislation could be passed and that he embodied U.S. failure, Donald Trump embodied their rationalized populism. The conservatives claimed that Obama wanted to take away their second amendment rights, as if they had the right to carry automatic war guns—which were so repetitiously used to murder innocent victims that Obama, in compassion for families of the dead, publicly cried with frustration about the failure to legislate against them.

Conservatives argued that limiting weapons would jeopardize their constitutional freedom of state's rights and individual power. When Trump's biases became absurd beyond these rationalizations, when his lies could not be treated as mere competition, the Republicans had found their quintessential representative. It wasn't enough to campaign on slogans of *no more Obama*, or *put Hillary in jail*. In the vortex of the Republican aggression, voiced in Trump's failed rhetoric, bias against the Muslims, Mexicans, and all immigrants manifested the same in the alt-right aiming for the power of dictatorship.

In contemporary times, what is the difference between a Republican lie and a Democratic lie? When Trump was judged unfit to become president because he lied about facts by using bias as causal, the conservative people he gathered around him advised him to say over and over that Hillary Clinton was unfit to be president because she lied to the American people. When Hillary Clinton talked about the logic of policy, her meaning was denied by that attack. The mechanism of the projection of aggression is based on the projector's entitlement to take an all-out aggressive stance, to do whatever he wishes. There is no cognitive logic only the falsehood that there are good people and bad: the good agree with you as if they are kin, the bad are dehumanized.

Donald Trump's unbridled social lies to influence the people played well when media uses public opinion surveys to separate popular and unpopular people. Indeed, the internet distorts data by tracking self-serving predilections and then presenting the most appealing articles first, as a strategy for selling ads. In an era of information peddling, self-professed bloggers use their number of followers as proof of influence. I believe this manifests an immoral use of so-called public opinion. In cultural regression, rationales based on followers often presents biased moral standards. There is a tendency for bloggers and polls to replace honest journalists in informing the population what to consider in determining policy. When Trump hired Roger Ailes to inform his campaign policies for winning the debates with Hillary Clinton he hired the man who innovated lies as news—as long as lies brought in more advertising revenue—and afforded Ailes the ambitious and sexual conquests he wished. That is one view of immature narcissism.

In a multiethnic, multicultural nation, it is necessary to compromise with others who are not kin, who have their own clan, cohort, or religion. With recognition of many sides of policy questions, compromise is a key to logical governance, pinpointing a plan best for all the people. Compromise is not winning an argument based on mutual bias or name calling, but on a clearly rendered alternative policy for each issue in its context.

If Donald Trump represents immature narcissists, fixated to unbound wishing, Hillary Clinton represented mature narcissists. She felt entitled to correspond through private email accounts and her blackberry to her elite cohort. She believed that those who were Secretary of State before her used their own private servers. Thus, she used her server to communicate with well-placed world friends to finance the Clinton Foundation:

"We believe that the best way to unlock human potential is through the power of creative collaboration. That's why we build partnerships between businesses, NGO's governments, and individuals everywhere to work faster, leaner, and better; to find solutions that last; and to transform lives and communities from what they are today to what they can be tomorrow" (About us: Clinton Foundation.org). Yet that ethos of mature narcissists does not insure maturity of their sources of influence.

The internet communicates with the top and bottom of the cultural hierarchy, from global to tribal culture. Smart phones are more vital than toilets in much of the world. Isis uses the internet to win converts, to present religious rationales for its policies, and to raise funds for its goal of returning global culture to fundamentalist primal eras. Jihad arguments are used to win converts through the internet, by promising to restore quality of life—or death—to those suffering from a cultural syndrome. Jihadists and others use the term *collateral damage* to rationalize terrorist tactics, saying that using drones for military strikes by someone thousands of miles away dehumanizes aggression.

DOES LAW REPRESENT CULTURAL MORALITY?

One way to approach the distinction between tribal culture and global culture is through the nature of the laws that bind membership at each level. Is national or tribal law more compelling? Since morality is more highly determined by the national laws governing a culture, we can judge the breakdown of cultural structure and the loosening of its bound aggression at each level, by assessing obedience to national law. In *The Magna Carta Betrayed?* Jed S. Rakoff, U.S. District Judge for the Southern District of New York, argues that misinterpretations of American national law have set the tone for moral regression (Rakoff, 2016). He says that the Magna Carta should be used in America to represent the habeas corpus rights of the individual: that is, the right for protection from titular leaders. He says that the Magna Carta holds that no free man can be imprisoned, exiled, or stripped of possessions except with lawful judgment by his peers.

But who are his peers? Is any person's identity lodged in tribal identity, religious identity, state identity, national identity, or global identity? Laws are ambiguous because they are based on words that change context in a dominant collective culture. Rakoff says prisoners at Guantanamo Bay were deprived of their own sovereign rights as free men. He notes that *Marbury vs. Madison* shows that American Indian tribes are sovereign nations whose people deserve writs of habeas corpus against imprisonment by an American sovereign. If U.S. states fail to finance adequate legal representation before the law where death penalties or long imprisonment are in question—or gerrymander votes, national law is defied.

Rakoff argues that the right of the imprisoned to have counsel representation is supported in *Boumedian vs. Bush* in regard to Guantanamo detainees. Assessing how George Bush over-road habeas corpus, by artificially declaring war on Al-Qaeda when rebellion and invasion were not in question, Rakoff cites the illegal hamstringing of the writ of habeas corpus in the 1996 case of AEDPA (Antiterrorism Effective Death Penalty Act). Avoiding habeas corpus denies the right for defense against the death penalty by states that limit the use of federal law by prisoners who may be innocent. Indeed, the increased difficulty in overcoming states' incarcerations of millions of black people in the 1990s on drug charges corresponded to the same politics: that is, their reluctance to let their decisions of death penalties be reconsidered by federal law. Yet the Supreme Court vindicated the states' approach.

EFFECTS OF RECENT AMERICAN HISTORY ON BLACK RIGHTS

During the prelude to the Vietnam War, black's rights were at stake in the Civil Rights movement. In many states, black people were denied their right to vote, deprived of equal rights to education, discriminated in bids on housing, unable to use public transportation as equal citizens, and barred from many public resources. Various black organizations demonstrated, defied, and some finally threatened force to gain their rights. As other subcultures entered the conflict, assassinations of President John F. Kennedy, Martin Luther King, and Robert Kennedy, the attorney general, were victims of unbinding aggression—bringing Civil War memories back to some. U.S. cultural integrity was threatened by assassinations and violence. National identity had always been suspect, because black people had never been treated as full citizens. Defeating Indian tribes and wasting black slaves were once sources of guilt and shame in the establishment of a new nation conceived in liberty. Pioneer liberty of individual identity was enforced by the relegation of other's identity to that of subservience.

Bigotry is rationalized as moral superiority by the dominant hierarchy. But those held inferior are living proof of the fragility exerted by force. In the 1960s, I thought America unbound its cultural aggression by projecting it into a war with North Vietnam, rationalized as mitigating Russian Cold War influence. If citizens cannot feel patriotic and proud of their national heritage; and if they cannot feel that most immigrants will become patriotic because of their rights of equal opportunity to survive and thrive, then as the poet William Yeats often said: *the center does not hold.* Holding the center is a feeling of moral binding in the face of conflict. That was how I felt serving in an Army Reserve Hospital Unit during the Vietnam War. I wrote letters for conscientious objectors, explaining the disservice they would bring to army morale.

The Emotional Cost of War discussed the effect of the Vietnam War on American drug use (Harris and Newman, 1986). The availability of heroin to calm soldiers fighting in a strange jungle escalated the post-war use of drugs. Black soldiers were statistically more likely to be killed in that war than whites. Wars do not go well when the armed forces who have to fight them do not feel they are protecting their people in moral ways. Social bias increases vulnerability to PTSD, because trust in authority lessens vulnerability. The Vietnam War was not clearly moral, which was one reason servicemen were traumatized to a high degree. Many servicemen used heroin to blot out their fear of death at any moment, exacerbated by feelings of cultural conflict at home. Bringing PTSD and addictions home led to more street violence, as drug-based crime erupted in neighborhoods that changed after the loss of the war in Vietnam.

With increasing drug use in the 1970s and 1980s, despite Nixon's war on drugs, crime rates and murders, especially unbinding in the inner cities where many poor black people lived, triggered Bill Clinton's 1994 crime bill. Private prisons were funded by the bill. In my clinical experience as a psychiatrist, heroin binds aggression and cocaine unbinds it. The number of prisoners vastly expanded; mainly black people were sent to prison if they had three arrests for selling drugs. Then, there was less street crime, and the murder rate among black men lessened. Republican pressure reduced stays on welfare to five years. While black women had the opportunity to be nannies for white families, black men had little opportunity to earn a living. A small segment of the black male population, some of them athletes, could join the mature side of white narcissistic culture. Gradually, the opening to many blacks to enter the mature narcissistic culture increased, especially as ads showed black people enjoying the good life.

FREUD'S RECONSTRUCTION OF CULTURAL DEVELOPMENT

I am not using the term narcissism in a pejorative sense but in the sense of freedom to pursue a self-serving individuated quality of life. To understand the effect of immature narcissistic leaders on the development of cultural syndromes let's consider Freud's developmental model. His view of cultural evolution corresponding to individual cultural development led to his concepts of character disorder, neurosis, psychosis, and cultural syndromes.

Each stage joins cultural and biological qualities of identity consciousness. Freedom includes the right to narcissistic individuation and collective religious practice.

The infant's symbiotic equation of other with satisfaction is a brain-based readiness to equate identity feeling with both need and avoidance of strangers, who cannot be trusted as need satisfiers. Freud viewed infant symbiosis as the basis of group psychology: like hypnotism. Archaic groups are like herds in which all members seek maternal need satisfaction and paternal safety provided by the alpha male. Yet in ambivalence, familiar kin may seem like strangers.

A distinct change in identity separates preoedipal from postoedipal development. The maturation of prefrontal cortex leading to reflection on experience objectifies identity, such that the latency child realizes he or she is a member of a larger cultural community, beyond an extended family. At that point the child can love his or her own self as a form of otherness, which becomes the base of mature narcissism. The child realizes that each person has a similar tendency of self-love and social objectification; gangs or clubs become collective groups representing objectified identity, in terms of shared interests, status, and roles. The advent of superego corresponds to an indwelling feeling of cultural observation and judgment, which allies with one's cultural identity. Those children who are highly devoted to a preoedipal identity projected onto them by caretaker needs for their own narcissistic satisfaction, may remain fixated to a more biased way of seeing others.

Freud found that in latency, especially in boys, a family romance fantasy replaces the preoedipal identification with parents/grandparent with imaginary forbears. Family romance fantasizing is a way of breaking free from the extended family. In that sense, raw oedipal wishes energizing oedipal fantasies change their latency aim by objectifying parents as also in the same cultural world they inhabit. The transition to the club or gang and the influence of a teacher or coach leaves a residue of guilt about distancing parents. Immigrant children in particular feel guilt about their maturing narcissism as they accept a new national or global identity in camaraderie with other members of their generation. But, immature narcissists continues to love and hate the internalized preoedipal parent.

As a residency director in psychiatry in the 1990s, I read thousands of applications of foreign applicants, most from Asia or East European countries. Personal statements about why they wanted to become a psychiatrist often included a wish to help their native country and their family. Proud to get good jobs in their field, most who became psychiatrists stayed in the United States. Learning to treat people of different cultures with compassion and objectivity, added to their sense of maturity and self-esteem, leaving behind a residue of guilt. Perhaps everyone is ashamed of leaving behind ones family, tribe, clan, or religion in order to individuate. In *On Narcissism* (1914) Freud says loving self is a normal development of the ego. He ends that paper discussing the ego-ideal as an internalized cultural sublimation measuring self-esteem by how well one's contribution to cultural progress is appreciated by others.

ALTERNATING STAGES OF BOUND LIBIDO AND AGGRESSION

In *Clinical Neuroscience*, I presented evidence for developmental stages alternating hemispheres, from left brain (LB) binding of pleasure processes to right brain (RB) binding of aggressive processes (Harris, 1986). Binding raw pleasure in infancy, Freud's reality principle precludes hallucinatory unpleasure: biological failure of identity. Next, the young child learns to bind aggression by learning to avoid unpleasant affects that signal abandonment anxiety. Thwarting the young child's pleasure of defecating freely, the anal stage binds anal pleasure with rules of control. Here, I view the ego-ideal *good boy or girl* as the LB side of the preoedipal superego; the RB side—the ruling superego—manages obedience, binding aggression by rules. Introjected family culture limits raw aggression by modulating affect expression. But defiant raw aggression in tantrums is available to regression from later stages. Young children can develop stable ambivalence; but immature children cling to fantasies of symbiotic feelings of "*mine*," as if owning the mother. I saw a mother responding to her three-year-old whining after a tantrum on the descending airplane: "You hurt my feelings by making me sit still." She answered: "Well you hurt *my* feelings by not listening to me!" Immature narcissists act out their symbiotic wishes and their young child mineness.

Tracing childhood development to alternating libidinal and aggressive stages, Freud described a second libidinal stage from 42 to 60 months. Self-stimulating the genitals (masturbation) eroticizes fantasy, yet the child has to contend with anxiety over body integrity. A state of fantasy thinking has to be kept separate from practical thinking. In latency, a second stage of aggression—objective autonomy—becomes bound by the superego as reflection tracks goals, results, self-esteem, and cultural status. Past memory becomes continuous as the hippocampus records experience, making it verbally and emotionally available to identity reflection. The indwelling cultural superego built from diverse identifications resides in reflection. Mature narcissists may act out their oedipal age and latency wishes.

THE PSYCHOLOGICAL EVOLUTION OF CULTURE

Contemporary neuroscience shows the brain in three modes: practical thinking deals with reality, fantasy thinking deals with social/cultural input, and emergency thinking shuts down both. When organic damage impairs practical thinking, the default mode domain (DMN) may shut down logical thinking and familiar identity. In organic syndromes, illogical fantasy during stress may break into the practical mode. MRIs in those disposed to mix logic and enact fantasy show DMN and executive thought mixing—like a waking dream enacting fantasy. People with chronic traumatic encephalopathy (CTE) may lose self-control, like people in group panic or psychosis. We'll be considering how cultural regression leads some people to enact their culture's syndromatic phantasies: unbound pleasure and aggression releases childhood wishes into action.

Comparing childhood developmental stages to cultural developmental stages, Freud imagined the primal horde experiencing one another in relation to the father/leader as symbiotes of group consciousness—like infants or cult members. Then, in the young childhood of culture, extended families formed tribes in which roles determined one's membership status. Aggression is bound by one's role. Taboo rules against aggression enforced by a leader were attributed to totem gods. In *Totem and Taboo* (1913), Freud reconstructed an oedipal scenario of brothers banding together to kill the father/leader. Envious of their father's entitlement they colluded to kill and cannibalize him, acting out a cultural phantasy. Then, living in chaos and unbound aggression with regrets and longing for return of the dead father/leader, the brother's acts left guilt in culture's wake. In years of practice I saw three patients, two men and a woman, who cut their victims in pieces and ate the pieces during fast psychotic regression to an archaic fantasy state. Searching for symbiosis, each enacted a phantasy of oral reincorporation of their identity (Harris, Jay and Pontius, Annaliese, *Dismemberment Murder: In Search of the Object*, 1975).

As tribes allied in what we may call cultural latency, they developed complex rules to control behavior toward kin and non-kin. Communities overcame innate fear of strangers through trade. As we'll see in the *Epic of Gilgamesh* in chapter 5, when trade brought strangers into their life, those who learned from them took new roles. Tribal roles diverged: hunters, gatherers, nomads, and settlers differentiated cultural roles for men and women. Freud said cultures that differentiated hierarchically into kingdoms, empires, and nations still maintained earlier forms of family, tribe, and totemic religion. Studying cultural syndromes shows how conflict develops when hierarchical cultures are dissonant to their constituent subcultures. Some cultures are more symbiotic (oral), some more rule-bound (anal), some more imaginative (oedipal), and some more utopian (latency).

SYMPTOM FORMATION IN INDIVIDUALS AND CULTURE

I described the neural developmental stages through which each stage of life passage corresponds to an endogenous trauma (Harris, 1981, 1986). Freud termed the shattering of infants' hallucinated reality wishes in passage to young childhood and the later shattering of oedipal fantasies *universals*. Passing to old age, people may reexperience a strong wish for heavenly satisfaction leading back to unrequited, unbound, unpleasure: hell again. The analyst and psychobiographer Phyllis Greenacre showed that identity passages in people with strong fantasy are incomplete (1971). Using imposture, for example imaginary identity, they create illusions. They do and do not resolve the distinction between reality and fantasy. Every young child develops transitional objects, wishful objects as real as a teddy bear. A predilection for fantasy and alternative reality develops in that way. Infant symbiosis is retained in the stage of young childhood in so far as words and gestures seem magical ways of fulfilling needs. Freud compared magical thinking to superstition in obsessive-compulsive neurosis—and to tribes lodging their wishes in totem god. A regressive stepladder may lead back down to earlier fantasies.

Just as the young child has to tolerate older family members having the power that he or she desires, tribal members cede power to the leader and the shaman who control the taboo with their rules. Taboos are the aggressive power invoked in objects. The young child must control raw aggression or lose identity standing in the family, with its attendant abandonment anxiety. Like defying taboo, disobedience may ruin social identity, just as a tribal member's defiance of a totem god obliterates living forever. The bedtime anxiety of the young child may feel like a prelude to the annihilation of consciousness. I was taught to pray: "Now I lay me down to sleep, I pray the lord my soul to keep. If I should die before I wake, I pray the Lord my soul to take."

In the oedipal stage of social imagination, earlier childhood wishes are vested in each parental figure. Four-year-olds may appear to be playing together, but it is parallel play; each child's imagination is his or hers alone, not a shared drama. Imaginative oedipal stories like fairy tales, promising forever and ever, form a family romance fantasy in latency in which derivatives of oedipal fantasies can be shared dramas. My daughter at six and her friend staged dramatic plays they made up together, performing for the proud parents of each. Freud absorbed Goethe's latency memoire of staging marionette shows with his sister Cornelia, as we'll see in chapter 9.

Many great writers like Sartre, Thomas Mann, and Goethe began their works in latency, remaining attached to the puppet shows they wrote. In latency, children identify with culture—their own and imagined forebears'; latency kids want to hear stories of ancestors to invigorate their own fantasies—and stories. Freud said that in family romance fantasies—pervasive in creative, narcissistic, and paranoid people—kids imagine alternative fathers: heroes (1909). We'll see family myths in the lives of Darwin, Mann, Sartre, and Goethe creatively reified as scientific or literary works; and how family romance phantasies may be acted out in cultural syndromes as heroic terrorism.

BIOLOGICAL SOURCES OF FREUD'S UNIVERSALS

Freud's universals correspond to developmental switches among brain network linkages. Psychologically, developmental passages amount to a new beginning of consciousness, like being born again in an emergency. We see this dramatized by Freud's Wolf Man case of an infant traumatized by observing a primal scene—the scene shuts down symbiosis (*An Infantile Neurosis*, 1918 [1914]). In universal passages to a new stage, Freud infers the prior stage becomes traumatically blocked. Then, because identity feelings are called into play when the brain must deal with unfamiliar episodes, anxiety signals trigger switches to fantasy in place of dangerous impulses. Given that the DMN activates in childhood, I infer that to consolidate new childhood stages the limbic cortex underlying cerebral cortex switches on the DMN to deal with difficult new problems. DMN thinking activates when the brain has to deal with an unexpected lack of needed resources or unfamiliarity. With signal anxiety, adults switch to fantasy many times a day: a state detached from action and perception.

Studying the development of neurosis, Freud concluded that fantasies from different stages of development can coalesce in creative regression during prolonged stress. Resorting to fantasy when an unexpected situation triggers severe anxiety, one may reflect on identity—deep feelings of being alive—before continuing the tasks of everyday life. Perhaps entering fantasy states—like prelude to orgasm—is an experience of a little death: suspending the practical tasks that underpin ordinary identity. During seeming instants of life and death converging, like Bowe Bergdahl's blackness in the prelude to deserting his Afghan army post, a person yields to the signal of potential nonbeing as the brain switches from practical thinking to acting-out phantasy.

The transition from signal anxiety to fear looms over ordinary reality. Joseph LeDoux, famous for researching the amygdala, distinguishes anxiety from fear, which uses different circuits (LeDoux, 2015). When a person has been traumatized, the amygdala blocks out the event, the unbearable affect it generated, and the imagery that portended danger. The amygdala can shut down practical thought and fantasy thought if they lead from intense prolonged anxiety to fear.

CULTURAL SOURCES OF CONFLICT

Like Freud, I imagine how cultural evolution selected which brain processes were critical to survival. In herds and primal hordes a leader endowed with higher capacity for male hormonal aggression promotes survival. In tribes, the leader pursues collective survival goals and the shaman reinforces myths—cultural solutions to problems in accessing safe strategies. The shaman binds feelings of love and aggression into collective myths by presiding over affect ceremonies that prepare the community for response or relaxation.

Freud's cultural theory pits innate conflict between the biological self that strives for pleasure and the cultural self that lives through otherness. He proposed that identity was generated by instincts, modulated chemically and hormonally, patterned by male and female identifications, and refined culturally. We'll see that the LB organizes the search for pleasure to satisfy metabolic and relational needs and the RB organizes the aggression-binding rules of culture that support individual safety through cultural morality. In his theory, cultural altruism gains when love is sublimated (bound libido) and when compassion replaces schadenfreude (bound aggression).

Anterior prefrontal reflection modifies midbrain instincts, manifesting the cultural evolution of human relationships and identity. Stanislaus Dehaene, Professor at the *Collége de France,* shows how the brain's prefrontal anterior *global workspace* organizes myriad networks producing psychological processes (Dehaene, 2014). Just as there are many sets of multimodal cortical neurons that respond to a variety of sensory stimuli, he investigated many sets of neurons that modulate identity, such as memory of cognitive and emotional experience, situational decision making, and empathy. Individuals choose from a wide range of culturally bound identity experiences during practical thought.

FREUD'S HISTORICAL PERSPECTIVE

Globalization was a cultural expression of modernity. For Freud, it was an ideal derived from the structure of the Hapsburg Empire, built on collaboration and diplomacy among nations. Its Prussian arm was accustomed to fighting a limited war to expand the Empire. Colonialism was rationalized as imposed cultural progress. Freud opposed a homeland for the Jews, because it would detract from their holistic cultural identity of working with others for progress. Many Jews converted, relinquishing ethnicity to enter modern culture as citizens, individual contenders for making their own impact. For individual identity, modernism meant healthy narcissism, the right to impact culture for personal gain.

Freud's cultural theories borrowed from the social science of his era, which held that individual identity mixed male and female qualities in ways that predisposed a person to characteristic cultural roles—active male and receptive female. Intellectuals of his time thought Jews were vulnerable to atavistic bisexual identity. Hitler distorted those ideas, claiming that evolution brought contending ethnicities into a fight for survival of the fittest; and Jews were atavistic advocates of holistic culture who should be eliminated, because they blocked ethnic evolution. Rakoff (2016) reports that between 1940 and 1965, lobotomies were performed on homosexual men as a legal procedure, "curing" what was considered pathological by the American Psychiatric Association. Freud fought such beliefs.

Freud was torn between allegiance to Viennese and German progressive cultures vs. disdain for Central and Ally War leaders who used religious morality to justify going to all-out war. Patriotism meant giving your identity to the nation despite the immorality of war. He thought volunteering often meant sacrificing to your father—submitting to your father-land. After the War he testified as an expert evaluating doctors who had sent men into combat who were shell-shocked or traumatized. He took Wagner Jauregg, his main psychiatric opponent, off the hook for using painful shock treatment to force soldiers to return to duty. When Jauregg later inoculated tertiary syphilis patients with malaria, producing some cures, he received a Nobel Prize. He in return was magnanimous in not blaming Freud in a suit by a patient Freud referred to a nonmedical analyst. After the War, Freud helped the needy get cut-rate analysis with both medical and nonmedical analysts.

Freud was narcissistically depressed during and after World War I. His three sons were at risk in the military, his daughter Sophie died of the postwar influenza epidemic, and her husband, Max Halberstadt was so injured in the War he could not care for their two sons: Ernst born in 1914 and Heinele born in 1918. Heinele, Freud's favorite grandson, died at age five from TB. Freud's cancer added to his infantile sense of irretrievable narcissistic loss and thus to his death instinct theory. Ernst came to live with Freud's family, taking the Freud surname, perhaps like Freud's revenant sense of his younger brother Julius. Analyzed by Anna Freud, he too became an analyst.

CONTEMPORARY NARCISSISM AND OLIGARCHY

My presentation of left and right brain types of cultural institutions in *How the Brain Talks to Itself* provides perspectives on cultural conflict (Harris, 1998). Using the identity parlance of self psychology we can say that primary narcissists remain immature and selfish, while secondary narcissists mature to the point of concern for others who are treated as self-objects. There, I made the case that the pleasure of making money energizes LB capitalist corporations in a fair playing field, while deploying aggression energizes RB bureaucratic institutions including the military, police, fire-department, etc., bound by regulatory laws for documentation and safety enforcement. When the private sector is monopolistic rather than competitive, capitalism is regressive. When bureaucracy is partial, favoring ethnic, racial, tribal, or religious subcultures, local states' rights prevail over national rights. When bias and greed form an impoverished, increasingly syndromatic subculture, unbinding aggression disrupts the subcultural equilibria and law loses moral fairness for the unified holistic culture.

Mature narcissists, like Mark Zuckerberg, George Soros, Michael Bloomberg, Bill Gates, and Warren Buffett, use foundations to infuse capitalism with future planning for cultural progress. They support an egalitarian bureaucracy that gives everyone a chance to thrive by sharing community resources with a capitalist private sector that is globally competitive, while giving small businesses a chance to grow. They form a group of individuals imbued with an exaggerated moral sense of their right to influence culture. Immature narcissists (oligarchs) like the Koch brothers disregard cultural concern for globally preserving resources. The brothers are famous for using their billions to reward politicians who favor their coal mining corporations.

Maintaining archaic control over women's rights to individuate, fundamentalist oligarchs want to maintain fundamentalist culture. Fundamentalist cultures that maintain their bias against all strangers are easier to control in a unified way, because leaders can maintain bias against any other culture—or religious sect. Perhaps some Saudi oligarchs fit that category. America's ultra-conservative oligarchs often espouse Christianity as a fundamental source of American ethos. Sometimes their ethos attracts biased institutions like the KKK. They interpret the U.S. Constitution as others might interpret the Bible as a concrete description of an eternal reality. Perhaps fundamentalist oligarchs predominate in Russia, and among nations whose economic power depends on commodity resources.

CLINICAL SIGNS OF CULTURAL SYNDROME FORMATION

Experiencing dissonance between cultural and individual morality uncovers incompatible systems. The pressure between cultural control and individual needs manifests in many ways. Military tribunals judge patriotism by the willingness to share the life and death of comradeship; religious councils determine the right of an individual's afterlife. Jihad—or Crusades hold one military or religious morality against the other. Global morality and global economic control is often at odds with both military and religious morality. Since morality is what a culture makes of it, media may side with religious morality, military necessity, or its own economic advantage. Disparity shows up in questions like whether free speech, free art, a call to arms, or carrying weapons is a legal right or a threat. When individual and cultural moralities collide, the dissonance may fuel public spectacle.

Consider Bowe Bergdahl, the American sergeant tried for desertion after he left his unit in Afghanistan and allowed himself to be taken by the Taliban. (McCarthy and Opel, *Bowe Bergdahl in His Own Words*, *The New York Times*, March 16, 2016) When his sense of military identity showed signs of faltering under stress, his father wrote him, telling him to listen to his own conscience. Five years after his capture by the Taliban, he was traded for five Guantanamo, Taliban prisoners. The conservative media blamed President Obama for saving an American who was a deserter. But the media did not report the event as a diplomatic attempt to get the Taliban to participate in the future Afghan government. Public identity conflict can unbind aggression when the government does not share its strategy with the people. In many media reports, Obama was treated as soft on terrorism, especially by Republican opposition.

In order to feel solid identity, a person has to integrate tribal love with patriotism. Bergdahl's failing identity integration, like the Tsarnaev brothers who committed the Boston marathon massacre, acted out his oedipal childhood fantasies of being a hero—in his case a patriotic hero. His journals, sent to his friend Kim Harrison, show that before walking off from his unit he was constantly doubting his capacity to resist the dark forces closing down his identity—unless he could find a heroic adventure that would release his freedom of self-expression (McCrummen, *Bergdahl's Writings Reveal a Fragile Young Man*, *The Washington Post*, June 11, 2014). Since he felt that his army unit commander was jeopardizing his comrades, he would be their hero and go to the higher authorities and report them, or he might go out on his own into China or the wilderness and live privately. Perhaps that is like the home-schooled identity he learned in latency, when his parents moved to a remote part of Idaho.

As his schizoid cultural fantasies took over, his practical intentions were muddled. He tried to internalize his father's morality of doing the right thing, but his father had also veered away from his own Swedish background culture. Those who knew Bowe, like Kim Harrison, who tried as a friend to care for him, knew his identity was unstable. She feared he might fall into psychosis or harm himself. Before joining the army he had been found psychologically unfit after a month in the Coast Guard, and he dropped his plan to join the French Foreign Legion.

From the military perspective, you live and die with your buddies who join you in the patriotic ideal of intersubjectivity; being members of the same military culture sharing an ego ideal. Your life is their life. In *Group Psychology* (Freud, 1921), Freud used the shared ego of the military to characterize the bond that men shared in their service. What was the army policy when Bergdahl was accepted for armed service in Afghanistan? Did they ignore the plain psychological truth that he was unfit to join the ranks of the military? After the fact the military wanted to show him as a deserter, which made the trade for Guantanamo detainees seem like a breach of national security. From that perspective he would be a traitor. Unable to integrate his morality with army morality, Bowe's vulnerable identity cohesion fragmented further.

President Obama and his advisors were trying to shut down Guantanamo, because it was an icon of the ill-advised wars and ill-advised torture. He had been in a double bind for all the years of trying to preserve the nation's ideals of a multicultural freedom for everyone. He was blamed for the aftermath of the policies that put the world into that bind—as if the Republicans wanted him to fail. When the vortex of unbound aggression that brought terrorism to the fore erupted into the body politic of American presidential campaigns, enmity between Donald Trump's supporters and mainstream Republicans broke out. Obama reminded America that Trump's identity-based bias aped the paranoid position that fundamentalist Republicans had used for years to get their way. Obama brought up Trump's earlier charges that he was not a citizen of the United States. Through fundamentalism, the paranoid position reverts to a symbiotic tribal level innate to infant symbiosis: love of kin and hatred of strangers. That archaic identity bias loses cognitive logic. The paranoid position lacks capacity for metaphor as a way of reflecting on differences among people. Consider the metaphor often used in psychiatric evaluation: "People who live in glass houses should not throw stones."

Sebastian Junger, the journalist who writes about cultural and social calamities also considered Bergdahl's relevance to the American psyche (Junger, 2016). In his view, betraying the "tribal ethos" of his military unit, Bergdahl put all the men in jeopardy, because fighting beside your comrades is a primal share in a group willing to die for one another. Junger says willingness to die for comrades is a male way of valuing the lives of one another equally, beyond individual narcissism. He says, men maintain survival in the duress of attack or traumatic emergency, while women maintain collective tribal life every day. He says the willingness to die as a hero saving a stranger shows the humanity inherent in bound aggression; and at a tribal level of shared identity, people are less vulnerable to PTSD. But the global inequality of money for meaningful jobs leaves people more vulnerable to PTSD and more likely to military suicide.

Junger notes that during the Blitz bombing of London in World War II, being forewarned about the coming necessity of caring for one another, there was no PTSD in the aftermath. Sharing morality binds aggression and leads to altruism. But when governments do not share security concerns with the people, being taken by surprise leads to more PTSD and less altruism. Moreover, globalism gives oligarchic sociopaths internet access to plant strategic lies in media in order to win world influence. *The New York Times* of August 29, 2016 reports Russia doing just that to keep Sweden from voting to join NATO. Russia's use of false information in becoming the chief world oligarchy added havoc to America's regressing cultural identity.

UNDERSTANDING THE DEVELOPMENT OF CULTURAL SYNDROMES

Perhaps the paranoid position is the last practical seeming response to severe stress. With a prolonged threat to one's life and the life of kin (and buddies are like kin), a regressive biological state intervenes. A severe stress response blocks cognition and transfers the experience of affect to the emergency response system, which may induce PTSD. Prolonged acute stress blocks adult reality and triggers regression to childhood fantasies attached to imaginary resolutions. Terrorism triggers fear and the wish for warrior leaders to face down fear with outward bound raw aggression, which is the regressive path Isis takes. Like their mythic progenitors—nomads on the Asian Steppes, Isis wishes to thrive off the riches of settled cultures. Unable to create an economy, they take one established.

Too many sociopathic oligarchies precipitated Isis. The breakdown of global culture inevitably cascades to unbound aggression in tribal, ethnic, and religious sectors. No subculture can withstand most individuals undergoing severe stress, trauma, loss of kin, and migration, without forming cultural syndromes. A syndromatic culture distorts developmental resolutions, which trigger individual regression when myths of family and tribal origin pervade individual fantasies with *phantasies*. Creative people grow up expressing cultural myths of heroes and heroines, sociopaths grow up joining gangs who steal what they want from culture, and empathic people bear witness. Family syndromes of retribution and restitution feed children's fantasies of becoming heroes and heroines of their family, tribal, ethnic, and religious groups. They transmit the epidemic—like the Boston Marathon Bomber Tamerlan Tsarnaev was to Dzhokhar Tsarnaev, and like both were to the American holistic psyche (see chapter 15).

Here is how it worked with the Tsarnaev brothers. Their Chechen father and Dagestan mother came from generations of Muslims who experienced so much PTSD, starvation, religious enmity, forced migration, and loss of their Caucasus homeland from the time of Stalin, that successive generations entered cultural syndromes of retribution and restitution. The brothers identified with their family's mythology, fantasizing heroic revenge and payback inspired by their ancient founders, and condoned by their religion. Each generation regresses further down their developmental ladder in search of recourse. At bottom in symbiosis, life is vested in collective others and death is beyond reflection.

Freud recounts how tribal taboos required warriors who killed their enemy to bring his head into the warrior's hut where he would ask forgiveness over months of loving the head as kin. In regression to the ambivalence of young childhood, death is not a fact, as much as a loss of cultural or family identity. Isis models itself on ancient Asian Steppe cultures that beheaded adversaries to intimidate other cultures whose resources were used for the conqueror's survival. Perhaps Jihad is a form of Islamic compensation for the social syndromes that were overcome in founding Islam. Dzhokhar Tsarnaev lamented for those he killed, like an ancient warrior loving his victims' heads.

The idea of cleansing war, perhaps like Mao's *Red Book* is part of the myths of new founders. Isis appeals to those who feel the sorrow and helplessness of lost cultural identity in past generations. The narcissism of Isis and Al-Qaeda warriors grows by destroying prevailing cultures that diminish their own. Internet technology allows them to connect to the myths that motivate those who have lost their sense of homeland and place. Chechens can go as mythic warriors to an Isis-controlled imaginary new state.

Syria's Civil War opened ripe territory for Isis. Syria was a crossroads of the Muslim world balancing Shiite, Sunni, and Christian religions and many tribes. After World War I the French and English carved Iraq and Syria out of the Ottoman Empire, which brought diverse tribes together. Bashar al-Assad used his tribe's Alawite military to dominate Syria; he added Hezbollah by cementing his military alliance with Iran and used Syrian Kurds as allies in dealing with Turkey. Allied with Russia and Iran, he defied the Sunni oil producers that were American allies. Nations bordering Syria wanted to keep their boundary and tribal influences within Syria. Turkmen near the Turkish border affiliated with Turkey. To maintain Al-Assad as their client, Russia made Turkmen an enemy. Like the Spanish Civil War, outsiders with selfish interests entered: sociopaths, revolutionaries, media, adventurers, and secret agents. With *Isis* acting as a rescuer through its archaic jihad control, people from Muslim cultures who suffered social syndromes were ready to join a new *Islamic State Iraq & Syria.*

THE VELVET GLOVE OF THE RULING 0.1 PERCENT

Money balances world trade; its distribution determines global economic reality. In dysfunctional capitalism, the global economic system's statistical indicators of trends become unreliable. With oligarchs in legal niches, the public is prey to their economic lies. When the oligarchs preempt a nation's resources, political myths are presented as explanations; cultural regression leading to unbound aggression is inevitable.

Those who want to maintain their bureaucratic position and who do not want their secure retirement blocked become belligerent, wanting to ensure their guns as talismans. Stoking their anxiety and animosity toward democrats, sanctimonious conservative leaders give them plenty of guns. The disadvantaged poor and those falling back see the 0.1 percent as the enemy. U.S. Tea Party conservatives and democrats like Bernie Sanders share that view, but disagree on reasons for the inequity. The poor live with raw aggression in their lives, with little hope to end their suffering in a better future; drugs may present an illusion of a future. Males often depend on gangs as they grow up; learning to express aggression in criminal ways can be justified as restitution.

During the prolonged decline in middle- and lower-class incomes and employment, conservative republicans claimed that the reason was national government deficit spending on Medicaid, food stamps, and high taxes for everyone. They defended states that kept taxes low and underfunded the state portion of such programs. If lowering corporate and individual taxes would so stimulate the economy that poverty would be all but wiped out, then it must be true that a rising tide lifts all boats, not just the seaworthy.

Many Muslim nations have oligarchs whose power resides in the capacity to produce oil for the global world's needs. But given global increases in fracking for oil and natural gas, OPEC is losing its clout and its alliances are threatened. The United States, Saudi Arabia, and the United Arab League as well as others have not yet sufficiently addressed increased ambivalence in their alliance given that the United States has little need for more oil. Russia's oil oligarchs collude in global decision making with other nations that have lost their equilibrium between leaders and their poor. World politics appearing to devolve on Muslim religious and tribal differences does not address the effect of oil as a commodity supporting the world economy. So the rhetoric about facing terrorism together, just as the rhetoric about facing Israeli Palestine conflicts obscures how oil oligarchs have used that conflict to maintain a modus operandi of projected aggression. These politics are hidden from common people, who rely more on religion and fantasy than on economic alliances.

BIG LIES RELEASE CULTURAL AGGRESSION

An influential leader lying for economic advantage is a template for collective projection of aggression. Think of Grover Norquist's immoral pledge not to raise taxes as a rationale for destroying the poor. The lie spreads like wild fire when terrorism is projected as the reason for cultural fear. The media uses unscientific polls about policy choices to spread rumors, as if fear and projected aggression were not what was being measured. Guns for suicides, murders, and public massacres in the United States are easily overlooked if they are not the work of terrorists—even if they kill far more people. Doctors prescribing powerful opioids rationalize their fees and pharmaceutical companies rationalize their influence on the doctors and the public, while more people die from the legal drugs than having their quality of life bettered. Legally available automatic weapons are more for domestic fear of losing their power advantages in the United States than dealing with attackers. If outsiders were as aggressive as those who own automatic weapons for protection, those guns would be needed in a civil war. It is a lie that if the crowd had guns massacres would not occur.

If public demonstrations become violent, many feel they need their own weapons. What if the police or military were to join the other side? Demonstrations against alleged police brutality, shootings at schools, workplaces, and Planned Parenthood, and bombs in public places trigger perplexed collective fear. In the cultural unconscious, the public wonders if demonstrations against police brutality can back-fire, which resonates with countries in the Middle East where the structure of culture broke down in revolution. Fearing refugees as if they were terrorists in disguise makes the innate fear of strangers feel real. Conferring citizen status on a stranger feels like danger to those who already have trouble surviving. Europe's problem is trenchant. Migrants and their offspring may outnumber those who represent their nation's culture. Right wing politicians court citizens who fear a loss of their status and promote aggression against migrants.

Just as individuals under attack may develop acute stress responses, a collective group emergency response may build momentum toward irrational solutions in which everyone is a potentially dangerous stranger. Cultural regression triggers collective PTSD, among groups already vulnerable from existing cultural syndromes. Terrorist activities can trigger emergency responses that add to the turmoil. Collective PTSD may lead to war. It's a question of whether practical thinking gives in to perplexity and fear and finally to a loss of individual and collective identity. That depends on leadership and whether media can deal with reality and calm irrational fears—like Franklin Roosevelt's inauguration words: "the only thing we have to fear is … fear itself" (Roosevelt, 1933). Yet fear itself, the innate emergency survival response, paralyzes rationality.

Chapter 2

The Cultural Regulation of Identity

Chapter 2 looks at culture from Freud's perspective of progress and regress. Culture imposes progress on an emotional framework of mammalian origins that is hard pressed to adapt to the discovery of technological and social power distributed unevenly among individuals:

> Human civilization, by which I mean all those respects in which human life has raised itself above its animal status and differs from the life of beasts—and I scorn to distinguish between culture and civilization—presents ... on the one hand all the knowledge and capacity that men have acquired in order to control the forces of nature and extract its wealth for the satisfaction of human needs, and, on the other hand, all the regulations necessary in order to adjust the relations of men to one another and especially the distribution of the available wealth. (Freud, 1927, pp. 5–6)

Chapter 2 also looks at individual development from Freud's vantage of libidinal pleasure as the aim of both men and women. One problem for contemporary analysts is matching Freud's male-oriented views of sexual identity with female views. Freud argued that culture regulates the sexual behaviors of its members using its own moral laws and education, often deleterious to the individual and to culture itself. Although he implied that culture used aggression to control individuals, a second problem we will introduce in this chapter is how aggression became part of his biologically based, dual instinct theory. In *Civilization and Its Discontents* Freud considered the enmity set up in cultural evolution between men and women (Freud, 1930, pp. 103–104):

> ... women soon come into opposition to civilization and display their ... restraining influence—those very women who ... laid the foundations of civilization by the claims of their love. Women represent the interests of the family and of sexual life. The work of civilization has become increasingly the business

> of men. … What he employs for cultural aims he … withdraws from women and sexual life … the woman finds herself forced into the background by the claims of civilization and she adopts a hostile attitude towards it.

In our contemporary world, sexual identity is composed of many qualities that are up to the individual to sort out. As Rosemary Balsam, a pragmatic, feminist analyst says, male and female bodies have separate and usually distinct features that are written in the stone of identity (November 23, 2015). That differs from the perspective of Darwin and Freud where each distinction comes from a common origin. The cultural regulation of identity continually changes over eras. Clearly enough, every male and female has both kinds of gender qualities, the residues of identifications that shape a person's identity tendencies. The resistance of women to being treated culturally as sexual objects, rather than their own free persons, has triggered modern culture to conceive identity as the result of what each person can make out of their own biological body, their social identifications, and their own brand of cultural regulations. It was wrong in my view, that women were considered by early analysts as people with a deficiency of penis: the projection of male perverts. Does every man have a degree of perversion to project?

How Culture Runs the Brain approaches identity formation with a neuroscience approach: the LB promotes subjective pleasure as identity in order to maintain metabolic integrity; while the RB objectifies the body to preserve its safety as well as social self-esteem to preserve cultural membership. Pleasure comes from within; safety relieves outside pressure. Without that distinction, discussions about the source of identity confuse biologically based subjective identity with objectified identity as a member of culture. Women influence the subjective roots of infant symbiosis because their hormones and oxytocin shape their caring attitude toward their baby, with ambivalence still in the mix. Women may promote biological symbiosis as the basic origin of subjective identification more than men. High testosterone in a boy baby promotes more crawling away from the mother, perhaps toward conflicted symbiosis.

DON'T SUBMIT: SUBLIMATE

The DSM V still denies neurosis. Yet Freud knew that for himself and other neurotics, false education about sex triggered childhood neurosis. He believed that cultural control of sexual experience induced neurosis through emotionally induced brain effects. In so far as sexual repression manifests cultural control without really helping people achieve well-being, such rules for behavior may become aggressive power ploys not necessarily obeyed by men in power. Studying archaic tribal cultures, Freud understood that rules about incest, for instance, were necessary to limit outbreaks of aggression that could destroy tribal unity. He realized that the repressed became a part of unconscious fantasy life. The other part, he thought, was the inherited innate residue of cultural experience.

Sexual conflict can animate individual fantasies; identity-organizing sexual fantasies play a key role in a young person's individuation. Yet, each person's emotional development is also determined by family and peer group beliefs and myths, which are largely shaped by those who hold economic and cultural power. After their procreative years, most people become more safety-minded, their bound aggression sometimes unbinds, reshaping their character structure. Freud's theories followed that trajectory. Sexuality was in focus for the first half of his life, aggression for the second half.

Growing up in Vienna from age four, Freud experienced a culture where Jews were given citizenship status and many could enter intellectual and professional life, if not the controlling Christian aristocracy. He wanted to be a member of that aristocracy for many reasons. One was that for a poor Jewish young man, finding enough money to get married meant that masturbation was his main sexual outlet. Freud and Martha were engaged for eight years until he was financially able to marry. Needing non-Jewish friends to wield influence for him, he simmered about the barriers he faced in becoming a Professor in Vienna. His ambivalence about being Jewish influenced his ideas about sex and aggression as basic drives. He could not understand his pervasive guilt about satisfying his drives. He searched for cause in his childhood experience and in past Jewish culture.

In fundamentalist cultures where aggression may be stronger than libido, even in the symbiotic stage, culture evolved such that men who were leaders seemed to feel that they owned women. That was the case with Shulem Deen whose wife was culturally controlled by their fundamentalist Jewish leader. Now in America, treating transgender people as being able to choose their preferred identity has become a bone of contention in many southern states where conservatives reign. Not only are conservative cultures tribal in that regard, but that seems to be part of their insistence on their aggressive rights to guard their women with protective guns. They are not happy with the liberal views that liberate women as equal partners in modern identity; nor happy about transgender females who have a penis, bringing their female identity into a female public bathroom.

SOCIAL ORGANIZATION BINDS AGGRESSION

Freud was immersed in Western civilization based on the Austro-Hungarian Empire diplomacy, which historically used compromise and great ceremony to deal with the surrounding aggressive powers. Freud's theory of progressive culture minimized Central Asian Steppe cultures. In nomad cultures, killing was a way to control the aggressive hierarchy that structured those civilizations. The anal, latency, and adult stages bind the aggression of cultural structure.

Every culture has collective myths, phantasies that hide—and sometimes reveal oedipal motivated killings that began new eras of culture. Ambivalence between cultural ideals and cultural control always simmers beneath the surface of cultural bonds. Myths impel cultural communication sustaining the internalization of cultural rules governing group process. Imbedded in myth is the historical residue of survival activities that gave rise to the leader's intuition of how to maintain the group's collective survival. Phantasies of entitled power bind social aggression, structuring rules for group survival, including ways of managing appetitive and sexual needs of individuals in their life together.

René Grousset's comprehensive *The Empire of the Steppes: A History of Central Asia* (1994) presents a history of Asia that shows the force of aggression structuring Eastern culture as much as sublimated love may have organized Western culture. Much of my understanding of Central Asia and Eastern culture is culled from Grousset's work. For three millennia before Freud, until artillery changed the balance of force in warfare, nomads from a wide array of Turkic and Mongol tribes occupying the northern steppes of Asia repeatedly attacked European sedentary cultures and what is now China and India. The traumatizing effects of these genocidal attacks grew worse as nomadic tribes became organized in empires.

With Buddhism's 2500-year Eastern spread from the Indus valley, nomadic warriors and their victims in settled cultures felt relieved to believe in reincarnation myths accompanying relaxation techniques for reducing severe stress. By 1200 years ago, Japanese Samurai warriors, following Buddhist practice, could relax after combat, with tea ceremonies. Takeo Doi, a Japanese psychiatrist trained at the Menninger Clinic has written a convincing analysis of Japanese behavior developing from archaic symbiosis, still at the root of Japanese culture (1973). He points out that Japanese culture is ruled by female symbiosis: everyone must act as if they agree with one another; tribal warriors are aggressive until they calm down with flower arranging. Perhaps it's a question of which came first, the structure of love (LB) or the structure of hate (RB). Freud thought that after the oedipal crime and chaotic fratricide, the youngest son—closest to the mother—led a more maternal culture of bound aggression.

Spanning hunter-gatherer, pastoral nomad, and agricultural cultures for millennia, nomad cultures stamped aggressive marks on cultural practices. Structures aggression in military warfare, carried over to bound aggression in hierarchical bureaucracy. Those prefeudal structures may be added to Freud's ideas about how totemic cultures developed context for family and kinship in the transition from hunter-gathering to agrarian life. Sedentary cultures had to learn how to bind their aggression within their own tribal cultures in order to respond to nonsedentary cultures.

The ambivalent balance between aggressive power and libidinal satisfaction in maintaining human survival seems innate to culture, to the individual, and to his or her genetically determined biological instincts. Without a source of emotional satisfaction, people lose desire to stem an aggressive vortex of lost social identity and individual death. The spread of religions via the Silk Road brought Buddhism, Christianity, and Islam into play in balancing aggressive and libidinal forces within nomadic cultures. The development of individuals aware of their cultural roots, in both sedentary and nomadic cultures, did not bring tribal, animistic-magic shamanism to an end. Spiritualism and naturalism remain group practices to this day, used to relieve autonomic responses to major stress. That was the only form of treatment my Jewish orthodox patient wanted.

Like biological evolution, cultural evolution adds new structures to existing ones, such that the old ones are never lost, continuing as a cultural unconscious. Evolving from herd mentality, kinship groups developed leaders whose experience determined survival for the whole group. In a herd, a lead animal develops more physiological capacity to deal with stress and can signal other herd members how to respond to survival threats. In mammals when would-be leaders fight after the leader dies, the physiology of the new leader changes, producing higher cortisone and epinephrine responses to emergency, and daily higher brain levels of norepinephrine, which maintains vigilance.

Much cultural organization is based on rationalizing death: one's own and others. As larger tribes clashed, shamans came to use ceremonial myths of rebirth to assuage individual remorse over killing. Buddhism introduced sacrifice as a way to bind raw aggression. Sacrifice alleviates guilt for killing by joining the warrior with the victim, to keep him alive in a ceremony of identification. Buddhists elaborated these themes into religious practices that brought disparate groups under a multicultural umbrella. Perhaps identifications with so many dead gives culture its quality of virtual reality.

IS MIGHT RIGHT?

Superego moral authority developed culturally in the structural organization of nomads: individual life or death depended on following orders within a hierarchy. Individual moral judgment manifests cultural binding of aggression. In cultural regression, moral judgment lapses toward raw aggression, but to the extent that civilization requires laws and moral sentiment to ground the laws, we may speak of bound aggression, as Heinz Hartmann did (1964). Hartmann, Freud's hope for American psychoanalysis, formulated how culture binds aggression with its structure of sanctions.

In nomadic herding culture, what we consider devaluation of women today was standard practice. Childbearing provided a way of replacing war death. Homeric and Asian epics in the first millennium BCE show kings possessing as many wives and concubines as they could afford. The fecundity and empathic capacity of women neutralizes some of the raw aggression that men used toward their survival. Cultural elements of those times continued in Freud's own time, as religious control of female sexuality maintained the power of the state. Fundamentalist cultures still attempt to maximize family size to increase their own tribal power. Yet as Margaret Mead noted, women were valued for agricultural work in gathering and farming cultures (1928).

Perhaps moral legitimacy through sons was a religious, divine right constituting the authenticity of economic power. With paternity questionable, family members were murdered to consolidate power. Thus, in the Jewish religion, maternal lineage was more reliable than male lineage. Morally bound religious practice provides regulations for secular power. Organized religions bind aggression by balancing charity for the poor, against a bureaucratic institutional power hierarchy. Ministers sacrificed some personal pleasure for the sake of the common good. Religious authority splits innate ambivalence between leaders and their shamans who administer moral binding.

Norman Davies's epic history, *Europe: A History,* contrasts Eastern and Western European history from the Ice Age on (1996). The feudal system of the middle ages in Western Europe, for instance, depended on the Church to sanction its structure of moral and aggressive authority in the rules for each class. The long legacy of techniques for leadership control of cultural aggression shows up in Machiavellian ideas about the use of power to maintain hegemony. Bismarck's *Realpolitik* strategy for increased colonial control in Freud's time continues as diplomacy. Diplomats still trade power over tribal resources. For instance, Kurds feel betrayed by American diplomacy in the Middle East.

THE EFFECT OF CULTURAL CHANGE ON FREUD

Freud felt that by learning what the greats taught, he would be on track to make generous contributions to a progressive world culture. But learning about Hannibal or Moses did not imbue him with knowledge of how aggression shapes the social world or its leaders. Yesterday's aggressive leaders become heroes to following generations who enjoy enduring fruits of their power. That knowledge remained a split off ego theme buried in Freud's amygdala and hippocampus, along with his infant brother Julius put in his tiny grave by the one-eyed doctor/mayor Freiberg—who seemed to control life and death.

In *Introductory Lectures on Psychoanalysis, Lecture XIII: Archaic and Infantile Features* of infancy and early childhood, Freud reveals a source of his phallic aggression: "I once dreamt in a certain connection of a person who must have done me a service and whom I saw clearly before me. He was a one-eyed man of small stature, stout, and with his head sunk deep in his shoulders" (1916, p. 201).

That archaic phallic figure, the symbol of both preoedipal and oedipal power locked Freud into agreement with Jung about the archaic origins of the mind. It became the persistent image of instinctive sex and aggression that Freud always pondered. He said his mother confirmed this was the doctor from his early childhood. In a dream added to *The Interpretation of Dreams* when his son was on the front lines in World War I, Freud sees his son in the guise of a child barely out of infancy, reminding him of how when he himself was that age climbing up to a cupboard to get sweets, he fell gashing his chin. His mother confirmed that the doctor who stitched him up was the same stumpy one-eyed man. His mother was seven months pregnant with his younger sister Anna then, Julius had just died, and been buried by the same doctor who attended Julius and Sigismund during the dysentery that had killed Julius (Stewart, 1972).

Freud presented the confusion of his early memories as he tried to penetrate its mysteries, as he wrote out his self-analysis to his friend Fliess from 1895 to 1900. Details often became the subjective core of a new paper or book. His image of stealing life from a baby goaded him to seek his source of guilt until he died. In that archaic mélange, his *Screen Memories* (Freud, 1899) paper shows him in the guise of a younger colleague, telling a screen memory of joining his one-year-older step-brother, John, in stealing flowers and life energy from John's younger sister, Pauline—soon after Julius's death. In *Screen Memories*, an old peasant woman (Monika) brings life-sustaining bread to all the children.

Food for later dreams, like the War-time dream of his son replacing him as he himself had been the young child reaching inside a high cabinet for pleasure was associated with Freud receiving unexpected money. It was also associated with an earlier dream in which Freud had confesses to his friend Fliess that he stole family coins to get Monika's affection—she was imprisoned for stealing from the family. Who was guilty? Freud always pondered the origins of cultural guilt and the reasons for aggression between Jews and Christians. Did Jews steal from Christians or Christians from Jews? He had lifelong ambivalent ambiguity about that.

I surmise that Freud's preoedipal cultural conflict also came to roost in his feelings about America. He distrusted America as a center of psychoanalysis, viewing it as believing its exceptionalism under god entitled it to worldwide financial hegemony. As we have seen, he concluded that Woodrow Wilson's Christ-like obedience to his father, a Presbyterian minister, saddled Germany with reparations that led to World War II. In their view, Wilson failed to learn the facts of the European world when he negotiated at the Peace Conference, nor would he trust advisors who knew the facts, because in his Christ-persona he identified with his father's grand rhetoric. Moreover, Freud's other writings on the development of civilization (*Totem and Taboo, Group Psychology and the Analysis of the Ego, Future of an Illusion, Civilization and Its Discontents,* and *Moses and Monotheism)* were over-determined by his own cultural conflicts.

OMAR MIR SEDDIQUE MATEEN: A FREUDIAN CONTEXT

The mass murder in Orlando's gay nightclub on June 12, 2016 was a salient event in the U.S. experience of enacted cultural and religious phantasies of terrorist retribution. Omar Mateen's act expressed the relevance of sexual identity in an archaic fundamentalist context. Many fundamentalist cultures based on tribal law maintain that homosexuality is unlawful and that men have the right to possess women. As Omar's father told the press, strict Muslims like himself believe that "God himself" punishes homosexuals. In this highly publicized act and using the copious data about Omar, I profile Omar acting out his version of a father god's vengeance. I condense too many accounts of his life history and actions to specify particular accounts.

Omar was a 29-year-old American citizen, born in Hyde Park, New York; eight years after his Afghan parents emigrated there from Afghanistan. He had an older sister 31, and two younger sisters 27 and 25. When he was five the family moved to Florida, where he went to school. Omar's father made a good living selling insurance and his sisters had decent jobs. Muslim and attending a mosque, they were all Americanized. The parents had at least one clash ending when the father reported his wife to the police for attacking him, while she said he threatened to kill her. Omar was known for his temper, attending a special high school for kids with behavior problems. As early as third grade, according to *The New York Times* he was "constantly moving, verbally abusive, rude, aggressive … much talk about violence and sex" (June 17, 2016). He had repeated suspensions often after making threats and becoming violent when he felt insulted as a Muslim and disrespected as an American. On one occasion he was met by his father who slapped him in the face in public before taking him home. In the aftermath of November 9, 2001, Omar told school-mates that Osama bin Laden was his "uncle."

From my psychiatrist perspective, these reports indicate a tendency toward ADHD (Attention Deficit, Hyperactive Disorder). Impulsive deficits highlighting aggression often indicate a RB disorder and can reduce the capacity to resist enacting fantasies of violence. Even as he was acting out his phantasy of being an Isis hero in the Pulse nightclub, he shifted his attention back and forth to Facebook. Pudgy as a kid, Omar bulked up on steroids and weight lifting in his teen years, wanting to become a policeman. In his late teens he would get drunk, blacking out in clubs. Overuse of steroids and alcohol damage the brain's capacity to control impulses. And yet, he ended up working nine years for G4S, a private British security company.

At 20, Omar made an identity change, changing his name closer to his father's name, Mir Seddique Mateen, by adding Mateen, meaning tough in Persian. They both went by Mir, which confers an aristocratic quality. He attended a Mosque regularly and traveled to Saudi Arabia in 2011 and 2012 on pilgrimage to Mecca. At the Mosque was a young man, Moner Abu-Salha, who joined Al-Queda's Syrian Nusra affiliate, becoming an American suicide bomber. During his shooting rampage Omar cited him as a *brother*.

It seems that Omar wanted to imitate his father's quality of heroism. His father, a fundamentalist Pashtun Muslim, ambivalently sided with the Taliban, advertising himself on the internet through an incorporated site: Durand Jirge Inc.—*Jirge* refers to Pashtun judicial prudence, bringing together contending Pashtun Tribes. He interviewed the American-supported president Ashraf Ghani in Kabul in 2015, and broadcast it in California. As a fake president of Afghanistan, wearing an Afghan military uniform, he suggested Ghani should be deposed by some hero. He was trying to further the goal of his Pashtun tribe to gain Taliban and Pashtun access along the Durand Line between Pakistan and Afghanistan. His self-advertising continued right up to the eve of Omar's terrorism.

Clearly enough Omar wanted to emulate his father's imposturous, tribal, persona as a heroic leader of Afghanistan. As an adult, Omar expressed tirades of rage about his father. Given that the father came from generations of Afghans who developed cultural syndromes from multiple wars and loss of their homes, Omar had an ambivalent, sadomasochistic relationship with his father. They were like two peas in a pod trying to escape from each other's violent, impulsive, phantasy, enactments.

A *New York Times* article reported on incidents of Omar's behavior while working for G4S (Opel, June 18, 2016, p. A, 14). When feeling attacked in his American Muslim identity, Omar spoke of famous Jihad heroes, as if he had a kin relationship with them. When Omar was fired in 2013 after an argument with a deputy of Sheriff Ken Mascara of St. Lucie County Florida, after he felt attacked and humiliated in regard to his Moslem identity, the FBI was notified. Sheriff Mascara is quoted in the TC Palm newspaper website:

"Omar became very agitated and made a comment that he could have Al Qaeda kill my employee and his family … he followed it up with very disturbing comments about women and … Jews and then went on to say that the Fort Hood shooter was justified" (Opel, June 18, 2016, p. A, 14).

Like his father, Omar used multiple Facebook accounts to publicize his political positions. After his first killings in the Pulse, for the next two hours he used the internet to call for the United States and Russia to stop bombing Isis, claiming he would otherwise continue Abu Bakr al-Baghdadi's vengeance. While holding hostages, in calls to 911 he referred to the Tsarnaev brothers, Boston bombers, as his "homeboys." He pretended attaching bombs to the hostages, then he searched the internet for reports of his doings.

Omar showed affection for his youngest sister by willing his house to her and her husband weeks before his planned terror act. Yet, he showed no qualms about leaving his second wife and three-year-old boy behind. That wife, Noor Zahi Salman, is a 30-year-old American-born woman, who left an arranged fundamentalist marriage years earlier. Her Palestinian parents, who moved to the United States from Palestine before she was born, had insisted she marry a Palestinian. Perhaps Omar had learned to be less fundamentalist with her than he had been with his first, Uzbekistan wife, who would not put up with being slapped into submission. Noor met Omar on a dating site; they fell in love, despite Omar's insistence on dominating her. He did let her learn how to drive in the months before his massacre. Perhaps she was still learning when she drove him to the Pulse nightclub. Omar called her during the massacre, wanting her to let everyone know of his triumph for Isis and Al-Qaeda over the United States. She agreed to check the media and told him she loved him.

ON ASSESSING TERRORISTS

An account of how Omar's identity rendered him susceptible to becoming a terrorist may provide signs for law enforcement agencies. Charles Blow wrote an op-ed. Piece in *The New York Times* on June 16, 2016 showing that Omar was home grown in his capacity to get and use assault rifles for his rampage. Concern that someone may officially work for Isis or another terrorist group may ignore so-called home-grown terrorists. But some American immigrant families, like Omar's, carry syndromes with them. The identity maturation necessary for forming patriotism requires new experiences of how to achieve economic and social survival—revising a superego is hard work. Cultural identifications, morality, mythology, and subjective identity arise in the milieu of childhood. To reduce the conflict of entering a new milieu, people must reorder their fantasy life and their capacity for new identifications toward new models of self-esteem. Omar's father remained imbedded in his Pashtun identity focus.

Given the nature of identity, we must assess many factors: conflicts arising in sexual preference; individual, family, and subcultural syndromes; prevailing national culture; media input; and organic brain state; all in their combined effect on individual identity formation. In that regard, Omar showed immature sexual affinity for both male and female roles, with a longing to be accepted by both sexes. Sex was mixed into ideas of violence in his mind. That dovetailed with his Muslim belief, strengthened by his father, that God will punish homosexuals. Omar believed men should dominate women to enforce obedience to Sharia law. His Muslim jihad phantasies, reinforced by propaganda in the media, reinforced his immature narcissistic wishes to be a hero. He projected his aggressive and painful identity feelings by blaming mainstream American culture.

Omar's father paved the way for Omar's phantasies of becoming a Muslim hero. Imitating his father's imposturous identity role in Pashtun and Taliban fundamentalism, Omar identified with Al-Qaeda, which showed him the way to overcome humiliation with self-importance. Humiliation leads to wishes for revenge; in this he followed his father's emotional example. Omar and his father both appeared to have a disposition toward ADHD, given their quick changes of political posturing, but organicity from alcohol damage and too many steroids could have pushed Omar even further into acting out his phantasies instead of controlling them.

Omar's cultural syndrome pointed to wishing revenge on Russians and Americans who revived the English colonial division of Pakistan and Afghanistan in the 1870s that led to the displacement of many Pashtun tribes. The Taliban is trying to unify those borderlands under their own terrorist control and Jihad. Omar searched for identity in terrorism as a political act of fundamentalist unification against the United States and Russia. While calling the 911 operator, he answered her question "What is your name?" with "I pledge allegiance to the Islamic State." This pledge of identity incorporated Abu Bakr al-Baghdadi's Isis imago as his leader—as he once incorporated his father, the aggressor.

AMERICA'S RESPONSES TO THE ORLANDO SHOOTING

What issues were on people's minds after Mateen's murders? First was whether the FBI should have kept him on their watch list. Second was whether those on the watch list should be allowed to buy automatic weapons—for example, weapons of war. Third was whether hate crime bias is directed against homosexuals either personally or from fundamentalist law.

Fourth was how much Mateen was "a lone wolf" and how much influenced by internet propaganda. Fifth, are immigrants longing for their past able to feel allegiance to a new nationalism, particularly if they are themselves subject to bias as immigrants. Another factor is how much paranoid thinkers seek social issues to dramatize their own identity problems and how much media seeks those same issues to excite the public.

Omar pledged himself to Isis as well as to Al-Qaeda because Isis is the catchword for American fears of an overarching terrorist enemy. He attacked American fear with threats of being part of a larger contingent of terrorists, in the same way his father played on his tribe's identity phantasies of united heroism against an enemy. Group fear triggers rumors and irrational behavior. Viewing terrorism as Isis or radical Jihad plots draws security agencies into fascination with external threats. Objectifying terrorist behavior treats identity as a choice of affiliation rather than a subjective process mediated by kin. U.S. policy in Afghanistan needs to recognize subjective Pashtun morality (Jirga). In judging terrorism and who the terrorists are we cannot let Al-Qaeda, Isis, or the Taliban become the defining enemy, without knowing the subjective attitudes, religious vicissitudes, and yearning for familiar natural milieus imbedded in cultural syndromes. Omar's father was perceived as both for and against the Taliban. But we deal with Afghan cultural identity through him. Pashtuns tribes developed Jirga councils over millennia. Afghanistan is not a place where living is easy or fertile, but a crossroads for trade. Nomadic Pashtun tribes evolved morals responsive to changing conditions: agreements are impermanent. Omar and his father altered their imaginary identity as events unfolded.

Omar tried to be American, but his identity would not cohere. His two wives each brought a different Muslim/American outlook to him (Uzbekistan, then Palestinian) that he had to formulate for himself by bringing it back to Muslim fundamentalist beliefs that did not have to deal with logic. His first wife was a secular Muslim Sunni who adapted better to America after her divorce from Omar. Uzbekistan Muslims tend to blame the Russians for failing to allow their secular religious practices. His second wife was also a Sunni Muslim who seemed secular. Sunni tribes, as political followers of Muhammad, tend toward dealing with modernism and global technology with political aims, while Shiite tribes see themselves as actual kin of Muhammad and maintain a tribal view of reestablishing past unity. These different views align with this book's contrast between hierarchical globalism and tribal fundamentalism.

Omar was a lone wolf whose attitudes and plans were inconsistent. Even after he decided to do the shooting, he still bought tickets for the next day to go to San Francisco with his wife and son because of an illness of his wife's mother. His lack of consistent logic, coming out of a conflicted identity, shows the breakdown of practical thinking and impulse control when phantasies of heroism preclude logic. Bowe Bergdahl had the same kind of identity breakdown precipitating his action, when he felt that his squadron leader was not dealing with the Taliban and he would have to take matters into his own heroic hands. I consider this as psychotic thinking without a total loss of identity integration.

Omar's behavior leading up to his shooting did not correspond to his stated attitude about homosexuality which was not biased, according to accounts of many acquaintances. We can turn to Freud for an understanding of homosexual ambivalence that leads to a paranoid position. When a father is violent but loving, the boy often wishes to be in his mother's place. In essence this sadomasochistic fantasy brings self-sacrifice into the relationship with the father; but wishing to be in the father's place also produces the sadistic position. As we go on in this book we will see the importance of this dynamic in cultural regression. Self-sacrifice for one's country or tribe is a core cause of heroism, risking self-identity while identifying with the father's aggression. Those same dynamics show up in Afghan culture: *bacha bazi,* sexual pedophilia with dancing boys among Pashtun War Lords, was ignored by Afghan Muslim authorities until the Taliban took their stand against it. American marine officers allied with Afghan War Lords who owned dancing boys, felt ambivalence and disgust. Politics makes strange bedfellows.

The American ideal of individual opportunity lends itself to creative technology and globalization, and most emigrants want the chance to participate in the U.S. culture of opportunity; but those clinging to their past culture are more likely to have phantasies about restoring it. Some families cannot give up their cultural syndromes when they move to a new nation. Omar's identity wish to be a policeman (he loved his NY policeman tee shirt as a kid) corresponds to his father dressing in an Afghan uniform pretending to a Pashtun takeover of Afghan politics.

Bias projects a person's aggression into the object of attack. Questions of bias affect large parts of the U.S. public: Are Muslims more likely to become terrorists? By claiming that all Muslims are Jihadist, Donald Trump stirred the fear of voters, satisfying his insatiable appetite for influence and reassuring him that his identity is intact—safe from humiliation. Projecting his humiliation onto Obama, he *fires him*—saying Obama should no longer be president because he was responsible for the Orlando shooting. Did Trump internalize his father's biases? That all others have rights is the backbone of democracy. True democracy needs mature narcissism and empathy for victims.

So, why can people buy weapons of war as a cultural right? Perhaps underlying this "right" must be a kind of terrorist phantasy that if their advantage in the cultural hierarchy were to be threatened, owners of automatic guns could pose a threat of their own. Just so, the Austrian regressives during the late 1920s of Freud's time loaded up on guns of war, while the progressives tried to help the poor. The regressives shared fantasies of using aggression to maintain their advantage. Let's not waste the cultural lesson: An Austrian Hitler hid his grandiosity in the politics of primal bias.

U.S. policy is inconsistent in negotiating with the Taliban about Afghanistan's government. That policy is not open to the public. The Taliban too are inconsistent about negotiating with the United States. As we saw in the Introduction, the CIA, FBI, Homeland Security, military command, and the State Department were not on the same page as the president and his advisors when influencing public opinion about how to judge Sergeant Bowe Bergdahl's war behavior in Afghanistan. The changing scenario of war and diplomacy complicates assessment of Omar Mateen's terrorism, just as it complicates the question of whether Bergdahl was a deserter who gave the Taliban an advantage. Just as the American public condenses Isis and Taliban in terms of understanding their differing politics, terrorists too lose track of their complex cultural war. Moreover, security agencies do not share the view of military morality involved in regulating policy and internet data. Edward Snowden's leaks showed the essential conflict among public opinion, and security agencies. How can people in a democracy judge the relevance of world events if they are incapable of dealing with cultural truths?

PROXIES, PATRONS, AND PRETENSE: INFLUENCING CULTURAL IDENTITY

In "How to Understand Isis," Malise Ruthven sheds light on Abu Bakr al-Baghdadi's internet influence on Omar's cultural ideas, providing a context of why the Arab Spring failed and how cultural regression led to civil wars in the Middle East (2016). When George W. Bush led the United States into destroying Saddam Hussein's dictatorship in Iraq, his advisors did not understand the fragility of Middle East power. Baathist Sunnis under Hussein controlled a mainly Shiite population. National and tribal sectarianism interacted with clannism, authoritarian patriarchal systems exchanging protection for group loyalty by the members, who receive sponsorship and "a share of the spoils." Thus, it leads toward supporting widespread terrorism. Ruthven treats Syria as the exemplar. He says that YouTube videos were used by local militias to advertise their service to patrons of militias, as patrons became the source of local influence. (Ruthven, June 6, 2016). Thus, sociopathic gangs enter civil war through new technological communication. But who become leaders of gangs using sectarianism to their advantage? Ruthven picks up on Fawaz A. Gerges's history of Isis (2016) that shows Abu Bakr al-Baghdadi becoming the leader of Al-Qaeda by transforming Baathist ideology into his own version of Muslim ideology. Using false self-genealogy he could fraudulently claim to legitimacy by virtue of descent from the Prophet's tribe (Ruthven, 2016).

Baghdadi was an idealized transference father figure for Omar portraying the heroic mentality of identifying through phantasy with symbiotic power, spread through the internet. We would not expect Omar or his father to be masters of logic. However, the developed world's approach to defining terrorism in terms of Isis-organized attacks also suffers from incomplete illogical approaches to evaluating potential terrorists.

In global culture, the wishes of politicians, lobbyists, and corporate leaders merge into a holistic economic influence of collective oligarchies. On June 24, 2016, Britain voted to leave the European Union. To David Cameron's dismay, the elite order of influence failed to hold the center of the global body politic, threatening global regression. Cameron, like many British leaders, including Tony Blair, was intrigued with the power of the elite leaders of the United states and Britain. Hillary Clinton too was a member in good standing of the contingent that felt the authority to bring down Iraq's Hussein. It is a product of undo narcissism, even a residue of colonialism that brought these leaders into a miscalculation of their global power, when so many in so many nations were left behind, while global technology of the internet opened their eyes wide to what they were missing.

Chapter 3 will explore how cultural influence evolved from innate, conflicting brain tendencies to form a multicultural hierarchy. We'll see how Freudian principles help us understand how institutional structures extended brain/mind process. Tribal love binds culture into the familiar biases of kinship; global aggression binds cultural structure into a hierarchy of subcultures and roles, using strangers to one another for what they can add to collective survival. When leadership loses its concern for holistic culture, the global cohort of narcissists join a new tribe of oligarchs; oblivious in their hubris to the lack of opportunity and suffering among the poor and middle classes. Consequently, cultural regression unbinds worldwide aggressors who act out their biases in terrorism.

Chapter 3

The Freudian Brain

Practical thought and default domain thought comprise two brain states that correspond to two kinds of cultural leaders: task-oriented and socio-emotionally oriented. Task leaders and shamans evolved to guide members' survival strategies. The leader makes cognitive decisions; the shaman makes emotional decisions. It seems the medial brain evolved as the social/emotional regulator and the lateral brain as the cognitive regulator. To address both kinds of survival problems, the brain harmonizes subjective feelings with outer physical and social reality. Humans' medial brains contain more recently evolved Von Economo neurons than any other mammal (Watson, 2006). Facilitating emotional communication, they link medial prefrontal network nodes with underlying limbic cortex networks facilitating response to motor task difficulty or sensory surprises. Boyatzis, Rockford, and Jack distinguish practical leaders from emotional leaders:

> … the division between task-oriented and socio-emotional-oriented roles derives from a fundamental feature of our neurobiology: and antagonistic relationship between two large-scale cortical networks—the task-positive network (TPN) and the default-domain network (DMN). Neural activity in the TPN tends to inhibit activity in the DMN, and vice versa. … The DMN plays a central role in emotional self-awareness, social cognition and ethical decision making. It is also strongly linked to creativity and openness to new ideas. (Boyatzis, R.E., Rochford, K., and Jack, A.L., 2014)

Chapter 3 sets out contemporary neuroscience findings of brain network functions that correspond to mental and social functions. I will point out correspondences between Freudian understanding and neuroscience understanding of modes of practical, stressful (DMN-fantasy), and emergency states affecting thought. Here, readers have a choice of studying the brain-related details, or staying with the mental and cultural conclusions. The Human Connectome Project portrays three kinds of thought that bring clustered networks into play during the assessment of experience in the context of both immediate and pronged survival needs. Hundreds of "small world networks" form the functional brain during development. That is why familiar identity—who one is—has multiple locations that overlap in the formation of identity within the brain.

Erez Simony et al. studied subjects whose DMN brain networks functioned in parallel while listening to a provocative story narrated over time (2016). The DMNs activated whenever long-term emotional memory was evoked. "DMN coupling strength predicted memory of narrative segments" (2016). The persistent maintenance of DMN during an experience of narrative that evoked emotional responses that were unexpected manifests the interaction of DMN thinking with practical thinking. DMN thinking, like fantasy, is recorded in multiple brain networks clusters requiring complex consolidation. The authors conclude that "DMN function has been linked to the recollection of past episodes and the stimulation of future episodes, theory of mind and tasks involving social content" (2016).

The 100 Functional Connectomes Project has developed a multicenter database of brain functions (Biswal et al., 2016). At 0.1 seconds, functional networks from multiple brain sources activate together without regard to specified anatomical circuits. In this way mental and cultural qualities shared by people present coordinated brain functions like those cited in the Simony article. Distinctions of sex and age are correlated in comparable temporal patterns of nodal network activations. Since shared cultural identity experience organizes community behavior, we may expect to find many attitudinal dispositions among peoples sharing functional brain characteristics. *How Culture Runs the Brain* will examine how shared cognitions and emotional attitudes build up socially and culturally into shared identity tendencies that may affect collective ongoing human brain evolution.

ABBREVIATIONS FOR MIND/BRAIN NETWORK FUNCTIONS

ACC: Anterior Cingulate Cortex: Medial limbic cortex detects LB error and RB salience.

AIC: Anterior Insular Cortex: Lateral limbic cortex distinguishes familiarity and novelty.

DA: Dopamine: Pleasure-linked neuromodulator system drives action toward goals.

DMN: Default Mode Network: A reverie state of social fantasy responds to novelty.

DMPFC: Dorsomedial PFC: The LB assesses goal progress, RB observes reality milieu.
LB: Left-Brain: This hemisphere organizes verbal, motivational, and pragmatic functions.
NE: Norepinephrine: The affect neuromodulator responds to novel perceptual challenges.
OFC: Orbital Frontal Cortex: Conditions social experience to LB pleasure and RB pain.
PAG: Periaqueductal Gray: Midbrain centers regulating innate instinctual responses.
PCC: Posterior Cingulate Cortex: Central limbic cortex that coordinates brain functions.
PFC: Prefrontal Cortex: Highly evolved cortex that tracks ongoing, overall experience.
RB: Right-Brain: This hemisphere organizes affect response to holistic reality experience.
TPJ: Temporal Parietal Junction: Accesses past social choices during present conflict.
VMPFC: Ventral Medial PFC: Matches LB motives and RB affects to self-identity needs.

The PFC above regulates VMPFC feeling and DMPFC reflective consciousness. Many networks participate in both task-oriented and DMN fantasy-oriented states, switching off and on depending on the task or observations involved as well as the kind of assessment needed to deal with conflicting information. Mind divides LB action preparation and RB perception of the real world. LB emotional networks regulate pleasure motivation, and RB medial networks regulate affects that block or release goal consummation. The medial parietal area—*precuneus*—provides motor memory of how to achieve practical social results by LB intentional or RB imitative actions: "how I did it." The temporal parietal junction (TPJ) provides emotional memory in LB words and RB images: "what it felt like." The temporal lobes' *hippocampus* provides associative long-term memory, LB declarative: "I remember how I did it," and RB episodic memory: "you taught me a lesson!" During identity frustration, we may enter prolonged DMN fantasy.

MEASURING EGO FUNCTIONS

Neuroscientists have many ways of measuring activation in functional brain networks: network nodes that form functional mental hubs. We'll explore how Freud's structural theories are consonant with their brain imaging findings. Replenishing metabolic resources and reproducing our species are LB problems of needs and appetites. Protecting our physical bodies from animals, nature, and one another—and our social identity from one another are RB problems. The brain solves these problems by assessing frustrations in need satisfactions and dangers to safety; and changing its expectations by fantasizing identity changes as trial actions, and by consolidating new learning.

Freud described conflicting brain-organized functions in terms compatible with the complex interplay of hereditary evolution, brain structure, mental organization, and cultural development. Learning from his clinical work with brain-damaged children in the early 1890s and from lab work on neural histology specimens he was inspired to explore the roots of mental tendencies in his *Project for a Scientific Psychology* (1899). In later experiences of World War I, he posed innate cultural influences as a life versus death struggle played out in every person and in every culture. Early on he believed aggression was part of the sexual instinct; later he thought it allied with the death instinct. He posed libido and aggression as cultural instincts pitting individual narcissism against cultural control.

Heinz Hartmann, an analyst versed in brain function, and one of Freud's inner circle, merged ego psychology with psychoanalysis, writing of bound and neutralized sexual and aggressive instincts developed for conflict-free ego and cultural functions:

"Freud has repeatedly stated that the ego works with desexualized energy. It seemed reasonable … to broaden this statement to include also other energies derived from aggression. … The term neutralization refers, then, to the process by which both libidinal and aggressive energies are changed … toward a noninstinctual mode" (1964, pp. xii–xiv).

CONSONANT BRAIN AND MIND STRUCTURE

Freud's *The Interpretation of Dreams* (1900) presents a brain/mind structured process that creates the manifest dream through a regressive and progressive phase. That work showed brain process becoming mental process. Each day we repress conflict-inducing wishes (LB), as their verbal intention enters the brain's REM sleep, making contact with the brain's verbal and image-associated wishes still held in the abeyance of earlier life fantasies. In the progressive (RB) phase of a dream, sensory image associations to the wish form the manifest dream. Yesterday's LB wishes accommodate to today's RB waking reality. As the poet Wallace Stevens said in *Sunday Morning*: "She is content when wakened birds before they fly, test the reality of the misty fields by their sweet questioning" (1956, p. 68). Every night, by consolidating new learning, brains adapt to yesterday's reality.

Freud's 1923 structural model: ego, id, and superego added culture to his 1900 topographic model of neural system structure: conscious, preconscious, and unconscious. Both models fit functional brain networks elucidated by neuroscience. Current concepts of functional networks, organized by *executive functions,* correspond to Freud's concept of ego functions. Executive functions are personalized cultural identity functions working top-down within the brain. Functional brain imaging shows them divided into left and right organizations. Mind and brain functions work together under the auspices of culture: both indwelling and vested in external institutions. Freud used instinct, drive, intensity, conflict, defense, ambivalence, individual inheritance, cultural and biological evolution, developmental changes, symptom formation, mental and organic syndromes, dream, fantasy, myths, associations, drug effects, and effects of psychoanalytic interpretations and reconstructions, to uncover structural links between mind and brain.

RB vigilance protect us from danger, but when LB needs are unfulfilled; the brain has to prioritize what is most essential to survival. I view fantasy as a DMN way of dealing with these conflicts, sometimes creatively. REM dreams, daydreams, and DMN reverie detach us from behavior and attention to the external world. A DMN state of mind corresponds to Freud's clinical use of free association in a resting condition, detached from any intention or critique. Like lying on the analytic couch, the DMN engages the brain's resting energy while the head is enclosed by an fMRI machine. The person is observed and vulnerable. Considering what mental processes are engaged and what are not during DMN activation, this book explores our core understanding of fantasy.

NETWORK NAMES FOR EGO FUNCTIONS

The mental relationships among co-activated cortical networks in the functional brain are plotted as graphs of activated nodes in a hub. It is not only direct neural connections that create mental functions, but also connected firing patterns among widespread nodes. Van den Heuvel and Sporns identified nodes in an overall mental *connectome* (2013). These nodes comprise a structural and functional coalescence of mental qualities corresponding to ego functions. The mind combines consciousness of the real world, the visceral inner world, and a fantasy world that is partly a dynamic unconscious, layered from childhood. The DMN's link to emotional episodic memory in the hippocampus adds to Freud's premise that fantasy uses unconscious sources of past memory. To the extent that culture is a projection of mental functions that structured social institutions, we live in a culture that evolved a virtual life of its own. From that perspective Freud viewed evolved stages of past cultures as corresponding to innate stages of childhood development.

Measuring the activation of executive brain networks, Marcus Raichle concluded that two prefrontal medial networks provide task management, that determine whether DMN functioning may be needed. The DMPFC reflects on all conscious data, the VMPFC brings visceral data to consciousness as feeling and affect states. In this way the DMPFC becomes aware of conflict. Investigating why the DMN uses so much energy, Raichle reasoned that it regulates survival by creating priorities for resolving unpredicted social contingencies or physical surprises in the milieu. Those priorities involve daily reconsolidation of cortical networks throughout the brain. He describes the brain as a predictive instrument. It changes its salience search in a diurnal rhythm process that uses the DMN as a source of new priorities (2010).

From an identity perspective, medial networks maintain self and other identity integrations. DMPFC provides detached reflections and observations about the social, emotional, and cultural world; and the VMPFC translates the autonomic regulation of visceral organs as *self-feelings*. A transitional cortex between neocortex and limbic cortex linked to the autonomic nervous system, the VMPFC regulates the internal biological milieu. Thus, the DMPFC provides contextual analysis of ongoing relational experience; while the VMPFC actively links visceral, inner body feelings to consciousness. The DMN reconciles outer with inner relational identity—what goes wrong.

The PCC is the DMN's most highly activated network. This large central area of limbic cortex is like a central clearing station for what other limbic cortex networks found awry. Processing the Rolling Stones', "I can't get no satisfaction," the DMN responds "But I try." The PCC brings needs and long-term memory together (Maddock, Garrett, and Buonocore, 2001). Its high activation in wishful memory suggests that it links fantasy with emotional conflict and resolution. When working memory does not solve needful salience problems, the limbic system's visceral functions trigger specified feelings of need or frustration in the VMPFC.

The brain, like a switchboard, connects clusters of networks that regulate specified mental functions. Many nodes are part of more than one network both at the neocortical and limbic cortical level. The VMPFC for instance is part of the executive problem-solving network, as well as the DMN network. Off/on switches determine practical, DMN, and instinctualized forms of thought. The dorsal PCC remains activated during executive problem solving; while the ventral PCC, closer to long-term memory sources, activates with the DMN when task management is deferred, as it is in dreaming and fantasy states. The PCC, as the central network (or set of networks) of the DMN, responds to challenges and threats to the expectations of satisfaction and safety. When behavior cannot resolve problems, the ventral PCC triggers inner directed emotional thought to evoke sources of unfulfilled need by mobilizing fantasy. Leech, Braga, and Sharp outline these PCC functions: "The PCC is a major node within the default mode network (DMN) and has high metabolic activity and dense structural connectivity to widespread brain regions ... it ... appears to be involved in internally directed thought, for example, memory recollection" (2012, p. 215). Dorsal and ventral PCC switch off on status as task management and the DMN alternate.

Pathological acting out can occur when the brain is organically impaired. Brain-damaged from RB injuries (CTE: concussive traumatic encephalopathy), a person may activate ventral *and* dorsal PCC cortex during a search for satisfaction. Normally inhibitory axons are responsible for switching networks. But RB inhibitory axons crossing the corpus callosal divide are more vulnerable to losing connectivity because they lengthen more than LB ones during development (Bonnelle et al., 2010). When fantasy and reality mix, syndromes can develop that lead to acting-out fantasies: think of some NFL players—or terrorists—who have been in brain-damaging fights and have taken drugs to the point of blacking out.

In addition to comprehensive practical and fantasy hubs, the brain has other hubs that activate together during ego functions such as salience detection, novelty detection, cognitive control, and executive control. En masse, these functional hubs "make up your mind." Communication patterns among functional hubs indicate where mental switch points occur, as in "I just changed my mind." Van den Heuvel and Sporns (2013) wrote a compelling review of hub connections, showing that the most densely activated hubs are the DMN and medial frontal networks, which together integrate identity aspects of the social brain. Graph theory accounts for consonant functional brain networks and mental structures; while anatomical centers and synaptic connections account for brain structure.

FORMING IDENTITY

The term "hub" comes from mathematical theory about how many nodes must be activated for a particular configuration to be functional. Equating network hubs with the conscious working brain, I propose that the medial social brain, synthesizing other with self, is structured by DMPFC and VMPFC executive identity networks that regulate consciousness and preconsciousness, respectively. While linked to the DMPFC, the VMPFC remains linked to limbic visceral data. Freud thought that perceptual neurons were closer to the outside world and feeling neurons were closer to the inner body core. Dorsal functions are more cognitive; ventral functions are more emotional. Thus the DMPFC is reflective and observant, open to cultural communication, while the VMPFC is emotional and inward, open to LB pleasure/unpleasure data and RB distress/relief data.

Vinod Menon, studying the neuroscience of cognition, distinguishes LB and RB hubs (2015). The LB maintains intra-hemispheric hubs consonant with its functions of verbally tracking appetitive consummation and metabolic replenishment functions. The RB contains more inter-hemispheric connections corresponding to its holistic approach; "better safe than sorry." Thus the RB uses controlling switch points to inhibit or release LB intentions. From a Freudian perspective this corresponds to regulating defenses.

Menon emphasizes the core role of the PCC in the DMN hub, given its access to voluminous information in the whole brain. Citing the off/on activation between the DMN and cognitive tasking functions, he deduces that switch points that activate or deactivate brain hubs are intrinsic to mental awareness of conflicts and their behavioral resolution when tasks are complex. The DMN links to autobiographical memory and emotional regulation, while the medial prefrontal cortex evaluates value-based decision making. Menon concludes that charting the connectome for how hubs interact, promises to clarify how the brain transforms its activity into ego functions. Dehaene's "global workplace" moves in that direction (2014).

Meditating on necessary actions, other nodes in the DMN hub are called into play. The medial parietal lobe called the *precuneus* provides awareness of the outcomes of past behavior. LB knowledge of how past events were carried out—"how I did it in steps," and RB episodes—"how it looked to me as I did it," need not be directly recalled through the hippocampus. Instead, the precuneus stores reviewed versions of past choices, which determine behavior except in unexpected situations (Utevsky, 2014). Another DMN node, the TPJ, references past emotional decisions. One can make characteristic decisions without remembering specific events and episodes. That is why the destruction of hippocampal memory does not destroy characteristic decision making.

The DMN's precuneus and TPJ organize past prioritized choices. With nodes in the hippocampus and amygdala, I think of the DMN as imagining self and other in myriad past combinations. Moran, Kelly, and Heatherton point out that the midline core of the DMN, including the precuneus, DMPFC, VMPFC, TPJ, and PCC, process *self* as familiar associations and *other* as unfamiliar ones (2013). Keep in mind one further DMN node contiguous with the VMPFC, namely the OFC (orbital frontal cortex). The OFC comes online early in life conditioning the infant's social responses to what is pleasurable and what is not, what is self and what is other. Its bias of loving kin and hating or dehumanizing strangers underlies choices and decisions. Superego morality comes later.

Our brain's capacity to predict efficient behavior is the glue that holds cultural groups together. When predictions fail, we widen our attention to all the details. Still failing, we search for new survival ground, like tribes migrating to new grounds when old ones changed. When nothing works, we may regress into an archaic, infantile phantasy mode for radical change—starting over. This is what brings revolutionary ideas to the fore. At times it is visionaries with cultural ideals, at other times adventurers with sociopathy or brain damage who become heroes taking action based on family, tribal, and religious mythology to bring about a new era.

CULTURAL CHANGE AFFECTS IDENTITY

Extrapolating from the idea that every young child projects feelings of helplessness onto a father leader who contends with the outside world, Freud considered cultural phantasies to be projections of collectively developed childhood imagoes, that is, inherent templates for identity. In his metapsychology, identifying self with others in consecutive psychosexual and aggressive stages of development, in unique ways during each stage, builds character. Clinically, he noted that stages of identification can peel away like layers of an onion during paranoid regression or melancholia.

Relationships, cultural status, and body states influence one's survival prospects. When threats and vulnerabilities mount, identity questions like "Who am I really?" "Where do I stand?" "What can I do now?" roil DMN fantasy. DMPFC reflective observation continually judges the role of VMPFC-self in mental process. Dissonant economic, political, and cultural forces cause emotionally painful visceral responses in feelings about the self. After collectively seeking comfort and help through religion, some turn to cultural phantasy (myth), hoping to change their cultural milieu. "Who are we?" "Where do we stand?" evolves further into "What do we want? We want it now!"

KINDS OF BIOLOGICALLY DETERMINED CULTURAL EVOLUTION

What determines cultural progression or regression in response to survival challenges? Over the long run, evolutionary changes in the autonomic nervous system enhanced our capacity to prolong attention before employing emergency responses. But technological breakthroughs, better agricultural production, and better weapons changed the survival balance among cultural groups and their allies. Prevailing cultural morality determines collective behavioral tendencies and modes of survival. Cultures on the losing end of a war may be traumatized and develop cultural syndromes.

Chronic cultural problems induce people to take on culturally specific patterns of identification with one another based on forms of love and aggression. Before looking at these cultural differences in anthropological or neuroscientific ways, let's consider how two branches of psychoanalysis understand ambivalence and identifications in different ways. Kohut's *Self Psychology* deals with idealized self-extensions, so that self-mirroring adds to the feeling of being exceptional. *Self Psychology* is a LB psychology, useful in treating narcissistic people. Melanie Klein's [RB] *Object Relations Psychology* dealing with good and bad objects, is apt for treating borderline individuals. Such people experience bad others as transference objects: projective identifications of internalized aggressive objects. Using the term *object* for others—or even physical parts of others, reveals that identity composed of self and others is objectified: never completely familiar.

The attempt to scientifically objectify the physical brain and the mental self is a major impediment, I think, in neuroscience research; because it implies that if we could account for every physical process in the brain we would know it completely. Not so (unless culture per se is included in brain structure). Internalized culture determines much of the flow of functional experience. In typical neuroscience research, when the functional self is explored via activated brain networks, subjects are told to focus on their own personality traits compared with others. That's a request to objectify the self as well as others. But personality is as much made of the self's LB desires, wishes, and intentions as it is of RB body gestalt and social status.

Observing feelings linked to others, the DMPFC judges the status of goals and salience. The DMPFC's quality of social observation is detached from the VMPFC's direct emotional experience of others. The DMPFC corresponds to Freud's *judgment* function; while the VMPFC is the decision maker, setting behavior into impulse. Given frustration, we self-immerse in the DMN hub's wishful fantasy (LB), until its limbic cortex (RB) signals anxious awareness of being lost in fantasy. "What was I thinking?"

EGO FUNCTIONS AS BRAIN HUBS

Just like Freud approaching the mental system as a product of underlying brain structure, social neuroscientists are puzzling together the functional network hubs that together form a mental system. Without mentioning Freud, many researchers support his findings. In my reading of their data, the RB always inhibits the LB's untimely pleasure-seeking behavior. Marcus Raichle established the DMN as centerpiece of the puzzle, around which networks fit ego functions. Its *dorsal* PCC activates during reality perception, but when novelty surprises, the DMN switches off the outside world, and the *ventral*

PCC activates an inner directed state linked to the DMN's hippocampal nodes of episodic memory—which corresponds to a Freudian narrative of fantasy associations.

Adding to this Freudian narrative, pleasure enters consciousness as the VMPFC assesses visceral feelings, preconditioned by the nearby OFC as social signs of familiar satisfaction established early in life. In *Minding the Social Brain* (2013). I presented evidence that the left OFC links to midbrain pleasure centers, from which dopamine conveys pleasure signals. Is it feeding time? The right OFC conditions novel social situations as threatening or not threatening. Thus, the VMPFC arbitrates value-based decision making based on the strength of motivation and vigilance. The pleasure of motor activity (biting into it) is delayed while a person determines whether the fruit has mold. Thus, the left VMPFC initiates thought as trial action.

When novel situations interrupt meaningful tasks, right and left lateral prefrontal networks (attention hubs) activate. The left hub uses thought to search for remediation; the right hub searches for sources of novelty *and* increases vigilance if the situation seems threatening. Although, the DMN switches off in emergencies or overly novel situations, people with traumatic brain injury may be unable to inhibit the DMN at those times. Researchers Bonnelle et al. and Sharp (2012) showed that when novelty is high, the right AIC (anterior insular cortex) and the right ACC (anterior cingulate cortex) switch off the DMN—for safety's sake. They showed the effect of traumatic brain injury on the switch mechanism. Too much RB axon damage is at best distracting, at worst allows fantasy-driven behavior. In *A Critical Role for the right fronto-insular-cortex in switching between central executive and default mode networks,* Sridharan, Levitan, and Menon show that the right AIC and ACC trigger the switch back to central executive processing when needed (2008), exemplifying Freudian defenses against denial.

HOW THE MENTAL SYSTEM WORKS

Bressler and Menon (2010) reviewed research methods and findings about how the mental system works. They graphed *edges* of hubs to indicate long axon connections of a core function. Their problem was how to discern the direction of input and output nodes in the hubs. The RB usually has inhibitory effect on the LB. Measuring an ego function like perception or cognition may include an activated network in one hemisphere and an inhibited corresponding network in the other hemisphere. Research errors may happen if investigators measure average activation of the same named dual hemisphere networks. Because reflection on experience informs and influences cultural development, let's consider what kind of cognition it represents. Bressler and Menon (2010) showed DMN suppression during cognitively demanding tasks. Here is how they sum up cognition:

> An understanding of how the human brain produces cognition ultimately depends on knowledge of large-scale brain organization. Although it has long been assumed that cognitive functions are attributable to the isolated operations of single brain areas, we demonstrate that the weight of evidence has now shifted in support of the view that cognition results from the dynamic interactions of distributed brain areas operating in large scale networks. (p. 277)

Following this line of thought, I propose that DMPFC reflection and judgment includes the self-observing quality of impacting experience in a profitable way. Freud's executive identity functions, top his hierarchy of ego functions. Thus, each person's narcissistic wishes and socialized cultural identity regulate a conflicted mental system. During identity conflict the brain uses the DMN to balance individual and cultural priorities. Leaders have to continually revise their assessment of necessary policies for maintaining the survival and well-being of members. So they have to regulate their own identity change, for example, their fantasies of pursuing a new course for those dependent on them.

As the brain generates thought and feeling, thought and feeling generate changes in the brain. DMN identity fantasy must trigger daily brain change to keep current with prevailing conditions. Fantasy works on the medial social brain to reorganize a person's ongoing sense of identity. By exploring *regression in service of the ego,* the cultural art critic and psychoanalyst Ernst Kris helped Freud understand how defenses against past wishes bind creative energy still attached to unconscious dynamics, making it available for present time creative exploration of new avenues to survival (1952). He said that creative change always included an attempt to bring about corresponding cultural change.

Like Freud, brain researchers think of the executive control of the mental system as composed of flexible plans and responses in usual versus unusual tasks. Cole, Braver, and Raichle investigated practical process as cognitive organization of task performance:

> A lateral prefrontal cortex (LPFC) region's activity was found to predict performance in a high control working memory task and also to exhibit high global connectivity … within and outside the frontoparietal network. … These findings suggest LPFC is a global hub with a brainwide influence that facilitates the ability to implement control processes central to human intelligence. (2012, p. 8988)

Let's compare control versus uncontrolled process such that the degree of stress involved in discrepancies triggers different network responses. Thus,

I suggest the reader review the chart of network function of the ACC and AIC. For minor interruptions in executive functions, the AIC processes cognitive discrepancies; the ACC processes emotional discrepancies. AIC relays unfamiliar cognitive data to the dorsal LPFC for salience processing; while ACC relays visceral and autonomic data to the VMPFC, which processes *emotional* conflict in verbal terms. When emotional and cognitive data both produce discrepancies, the DMN is brought into play. When disturbing events cause visceral discomfort, the AIC switches off the DMN in what could become an emergency. Stress signals evoke a more complete limbic response to failed salience.

FUNCTIONAL HEMISPHERIC DOMINANCE

Freud thoroughly explored the neuroscience of his day. In his *Project for a Scientific Psychology* (1895) he concluded that perceptual neurons screen out over-stimulation, because the outside world has so much force that each modality can only be sampled. He thought that neurons in the interior of the brain only had to deal with the biological force of other neurons. Now we know that neocortex is multimodal; each grouping of cortical neurons specializes in processing different combinations of modalities. The RB processes the multimodal external sensory world specializing in visual and vestibular modalities; while the LB adds kinesthetic sequences to its specialized admixture of modes.

Are individuals genetically determined to favor one hemisphere over the other? Ninety-seven percent of people have LB speech dominance, but that could be a product of cultural influence (Nielsen, 2013). Does the use of language to increase social and cultural communication and to record historical events and the myths that accompany them affect hemispheric dominance? It is easy to fabricate cultural reality with words, such as how interest and political groups and their lobbyists interpret the constitution to pursue power.

RB holistic perception binds or releases aggressive responses. It integrates visual form and motion with vestibular up, down, and sideways sensitivity and perceives the direction of sound sources, all in the formation of three-dimensional physical reality. As Freud noted, skin modalities of touch, pain, itch, and vibration also integrate physical boundaries of the body. We know that bats' sonar and sharks' capacity to sense electrical stimulation in other animals enhances their sense of salient reality. Eat that Seal there!

Mirror neurons manifest appetitive intent. Manifesting the LB focus on satisfying appetites to maintain metabolic integrity, LB insular cortex (AIC) responds addictively to familiar satisfiers (Damasio, A., 2003). The LB integrates sources of human action. We use kinesthesia to make gestures of lips, tongue, and palate to form consonant sequences, speech signals for social communication about our action plans. We use inner and outer language to form muscle sequences to grasp tools or make gestures of intent.

The LB processes pleasure and unpleasure experience; the RB processes affect pain and its relief. In suspense, both hemispheres explore the unknown, sometimes with anxious exhilaration, gambling for example. Will pleasure or hurt be an outcome? Until the outcome is known, the autonomic system—the phasic parasympathetic and sympathetic systems—push some of us full steam ahead. When novelty indicates possible danger, vigilance triggers anxiety and retreat. If retreat is impossible, anxiety gives way to fear and stress. If stress or physical danger threaten life, limb, or kin, a traumatized mind can become temporarily useless.

FANTASY DEALS WITH DIFFICULTY, LIKE DREAMS IN SLEEP

Freud had a mature narcissistic mind, for example, he established his own conceptual framework for dealing with the world and a determination to work with others collectively to change culture if necessary. He had that in mind when he spoke of men sublimating their love for one another for the sake of cultural progress. Shared creativity brings us to consider how fantasy helps leaders deal with the world when individual and cultural problems conflict.

What can we learn about the psychological and neural organization of adapting to the unexpected? DMN is triggered by a RB failure to find necessary salience in the daily search for satisfaction, giving rise to a state of daydreaming and self-concentration, while shutting down practical thinking (Fox et al. and Raichle, 2006). Perhaps the DMN as the brain's mode of fantasy thinking begins the process of new learning necessary to deal with a lack of practical salience, a process that continues during REM sleep. Let's compare REM (rapid eye movement) dreams with the DMN function. Ellman and Antrobus used cognitive and Freudian approaches in evaluating the role of REM dreams in survival strategies (1991), and Ellman with Weinstein researched the effect of REM deprivation (REMD) on people and other mammals. REMD over multiple nights causes increased REM in subsequent nights and more affect expression in subsequent days. REM dreams must be a prelude to working out identity problems. Ellman quotes Harry Fiss: "We dream in order to focus … our attention periodically on what troubles us the most, perhaps even to work out some kind of solution" (1991, p. 311).

Deep slow wave sleep (SWS) is restful, as the plasticity of the brain's synaptic connections consolidates new learning from the day before. Waking from SWS, subjects report thinking from the previous day, including the frustration of being unable to solve problems. But waking from REM sleep they report imagery lacking reflection. I surmise that REM sleep continues the default mode of daydreaming, which is a separate state of thinking that is private and detached from action, just as REM sleep is detached from motoric intentionality. In the transition to the *tonic* REM state, thought is closer to the SWS mode, but in the *phasic* excitation of highly activated EEG waves during REM, muscle is deactivated by POG (ponto-geniculo-occipital) inhibition of spinal cord motor neurons: excitation has no outlet. Phasic REM may then correspond to the switch to the DMN mode of fantasy thinking triggered by difficult experience.

The evolution from reptiles to mammals had a major impact on human brain plasticity. Yu Hayashi et al. traced the evolutionary origin of mammalian mice cells that switch SWS to REM sleep (2015). Switching activation and inhibition of these cells is regulated by the nearby PAG (periaqueductal gray) and the Locus coeruleus (Hayashi et al., 2015). Thus, the control switch among modes of sleep is strategically located between instinctual sources, drive sources, and autonomic system sources.

When problems go beyond practical thinking into stress, REM sleep comes on earlier. I surmise that wishes in response to disturbing thoughts in SWS trigger the switch to REM sleep. Ellman and Weinstein (1991) report that REM deprivation causes daytime excitability: readiness to seek new salience and new initiative—even in depressed subjects. REM fantasy then, like DMN states of thinking about identity and identity change is a way of dealing with difficult problems that require long-term solutions, or which appear to require some basic changes in life style. If there is little hope of changing an overly stressful life situation, a person becomes anxious and may become depressed. REM deprivation can even evoke psychotic experience, with a loss of reflection and reliable identity resources.

During trauma, practical and fantasy thinking are blocked by innate emergency responses, consolidation is blocked, and regression to earlier stages of self-development releases primitive modes. As we go on we will distinguish between problems of daily life and extended problems of identity change. Daily changes trigger brain consolidation of new learning during SWS; while identity changes effect the interaction of multiple brain networks. Raichle showed that human evolution of high brain energy maintains plasticity of response to a changing world (2015). To adapt cortical neurons in response to daily exigencies, and especially core identity problems their pathways and glia use more energy than our brains expend in conscious mental work. Neural updates occur during night-time sleep cycles. Infants spend half their day and night, before and after birth, in REM states organizing their holistic working brain. Given massive turnover of neurons in the formation of ego functions out of the blooming, buzzing, confusion of primary and secondary cortical organization, and with apoptosis (destruction) of less useful or overly useful neurons, we can connect REM and SWS with the immense plasticity of the brain.

In chapter 3 we have examined how collective cultural syndromes can bring individual regressive fantasies into mythical modes. In situations of collective survival, stress, and loss of familiar identity, where kin are trying to stave off trauma, depression, or psychosis, individual DMN fantasy regresses into collective phantasy, holding archaic modes of culture in thrall within religious and tribal feelings of symbiosis. In trauma, sleep deprivation in both modes triggers a waking psychotic state. Dzhokhar Tsarnaev experienced intense dream-like states during the days before bombing as his expectation of survival diminished. As Bowe Bergdahl prepared to leave his squadron, he spent days and nights of a dream-like state recorded in his diary, describing the psychotic blackness of losing self, dying, or entering a new world. Omar Mateen was in that state during his prolonged night of shooting. The archaic heroism contained within wishful phantasies, typically brings a person back into the preoedipal mode; self and other are experienced as symbiotically united within the frame of merging with god to restore culture. Chapter 4 explores the passage to entering the unknown, where trauma may end the life one knows.

Chapter 4

How Mind Enters Trauma

Collective trauma has a double-edged effect of ruining individual identity and rendering whole communities syndromatic. Chapter 4 describes neural, psychological, and cultural mechanisms of these effects. Freud puzzled over the question of whether affects can be unconscious, while Jaak Panksepp and Mark Solms who have taken the lead in formulating neuropsychoanalysis as a science suggest that affect is preconscious, only requiring representation to know itself. They argue that an intrinsic, midbrain, affect repertoire determines what we can know about our feelings of self. Chapter 4 compares their formulation of consciousness with Freud's, and it explores the thesis that some experiences are so traumatic that any association to what happened feels like a repetition of affect so unbearable it must never again find conscious representation. According to Panksepp, "the primary process level is not 'unconscious' if one defines consciousness as the ability to have internal *experiences* ... by itself the foundational level may not be able to be 'aware' of its own consciousness. ... The primal level can only mediate *anoetic* consciousness—experience without knowing. ... We call this level of experience, affective consciousness" (Jaak Panksepp, 2011).

Contemporary neuroscientists have investigated the mechanism of a loss of conscious experience during fear induction. During traumatic fear induction the amygdala blocks the hippocampus and its interactive equilibrium with VMPFC self-regulation, which inhibits practical thinking (Koenigs and Grafman, 2009). It evokes a hypothalamic cascade of hormones that block hippocampal reception of unbearable affect experience. Suppression of the VMPFC leaves the DMPFC without its main source of motivational and affective data, which accounts for the sense of detachment that occurs during trauma, when practical thinking beyond perplexity becomes impossible, because of the hormonal effect on the phenotype-dependent function of VMPFC's processing.

When the cultural milieu offers no social remediation, fear conditioning cannot be set aside emotionally. When PTSD becomes syndromatic, any reminder, any image of the disaster, triggers the same right amygdala response of hormonal blockade. Spoormaker et al. showed that the extinction of hippocampal fear conditioning may occur in slow wave sleep (SWS), while extinction of unpleasure and painful affect may occur in REM sleep: "PTSD patients show a failure to consolidate extinguished fear … associated with hypoactivity in hippocampus and ventromedial prefrontal cortex" (2014, p. 1547). Practical thought which occurs during SWS blocks fear, fantasy thought which occurs during REM sleep blocks anxiety, but disturbed sleeping in severe and persistent trauma can block the brain's adjustment to trauma.

Chapter 4 considers whether innate cultural instincts lead inevitably to conflict. Gonweil and Jovanovic (2015) found that latency and pubertal children are very susceptible to PTSD, when growing up in a milieu of mass murder and other trauma. They are more vulnerable to entering collective syndromes than adults. Girls show more self-blame and higher rates of PTSD than boys—who more often project their aggression outward in wishful revenge. Stable age-related fantasy as a remediation of emotional hurt and stress is less consolidated when Prince Charming does not live in the neighborhood.

Jan Van den Stock et al. support this hypothesis from a brain network perspective (2015). The authors show that when witnessing a violent confrontation between two people, the observer will identify with the aggressor or with the victim depending on the observer's traits for aggression or empathy. When the aggressor beats the victim the observer's left centromedial amygdala activates the aggressive identification trait; but identifying with the victim activates trait empathy in the right anterior cingulate cortex (ACC) and right anterior insular cortex (AIC) linked to the right amygdala's basolateral nucleus. Those patterns fit our dual hemisphere model. Clearly enough, biased observers fit the narcissistic, aggressive trait category, by mobilizing their LB urge to get rid of people who threaten their pleasure; while empathy for the victim mobilizes RB networks.

BOTTOM-UP AND/OR TOP-DOWN

Let's reconcile Freud's top-down mental system with Panksepp's neuropsychoanalytic, bottom-up view of several instincts conveying affects. Panksepp says that affects are neuromodulated by *secondary process* neuromodulator drives, neurohormones, and limbic cortex on their trip to representation. Adding a *tertiary process* of abstract and symbolic mental regulation, he views consciousness as an innate set of affects, processed in preparation for being known in a cognitive context. Panksepp argues that early mammalian evolution provided instinctual sources for all the emotions we know (2010).

These sources of self-consciousness are preorganized around midbrain periaqueductal gray (PAG) in centers that link to brain sources of pleasure or fear, accounting for Freud's aggregate instincts of libido and aggression. Like Freud's sources of libido, Panksepp's pre-organized instincts of seeking, lust, play, and maternal care are innate sources of pleasure driven by neuromodulators and hormones. He sees no data basis for the James/Lang hypothesis that affects arise *de novo* in response to experience. You do not feel fear *because* you run. He sees secondary process drives enhancing instincts through subcortical conditioning centers (including the amygdala); and tertiary process cortical networks functioning to produce the mental abstraction and symbolism that attach to human cognitive processes as regulators. In his view, external stimulation of instinctual centers evokes the same affects any human experience does.

THE BRAIN PROVIDES CULTURAL CONTEXT

Panksepp and Solms lack a worked through cultural theory for how collective emotional life impacts the individual generation of affect. But simply put, every instinct has a social/cultural link. Culture provokes instinctual response. Panksepp and Solms link feelings of grief and loss to ACC, the limbic cortex that underlies medial prefrontal response to social and cultural conflict. The ACC equates self and others' emotional pain in a way that promotes empathy and social bonding, and it modulates frustration arising during dissonance between the cultural communication of morality and innate morality.

If Panksepp's *BrainMind* consciousness is a biological approach, Freud's *MindBrain* is a cultural approach. Here is Panksepp's full set of innate affect organizers:

1. The SEEKING system is a rewarding emotional system motivating familiar pleasure.
2. The RAGE system protects resources, avoiding restrictions on its SEEKING greed.
3. The FEAR system triggers fight, flight, or freezing, to avoid trauma or destruction.
4. The LUST system, hormonally driven even in infancy, matures throughout childhood.

5. The CARE system of maternal instinct instills empathy in both females and males.
6. The PANIC/GRIEF system regulates social attachment to avoid emotional pain.
7. The PLAY system socializes by facilitating displays of all instincts to the pack.

Solms and Panksepp describe a *BrainMind* bottom-up construction of conscious, preconscious, and unconscious layers: the id is conscious, the ego unconscious, with the limbic system still in between (2012). From my perspective, both Freud's and Panksepp's models correspond to the autonomic system's hemispheric divide. I view appetitive discharge as a LB function; and affect response to the objectified sensory world as RB. Neuroscience research confirms that the RB limbic system binds disruptive affects in order to go on processing a difficult reality; or under extreme circumstances it escalates, from triggering vigilance anxiety to fight, flight, or freezing, sometimes the prelude to PTSD. In trauma, when fear, panic, and rage discharge together they trigger amygdalar discharge to hypothalamic centers with a hormonal cascade that induces regression to a reptilian under layer of brain evolution. Cascading hormonal, neural, and genetic changes in phenotype disrupt practical and fantasy modes of thought and change autonomic brain responses to a more archaic pattern. In practical thinking, the RB binds aggression, using empathy for others physically or emotionally hurt; the LB coordinates pleasure processes.

REDUCING SUBJECT/OBJECT AMBIGUITY

Imagine the elusive enigma of identity. Because self feels unified in identity—except when it splits pathologically—subject and object tend to be equated in many theories of mind. Panksepp and Solms ambiguously treat affect as the only source of consciousness. But, the quality of subjective and objective experience is different, only merging when satisfaction and safety relieve conflict—synthesizing the nirvana feeling of complete identity. I view appetitive motivators as subjective signals of metabolic need, distinct from objectified response to the social and cultural milieu and to physical reality.

Like Raichle, I think the brain is a predictive instrument responding to unexpected change with a set of innate, culturally determined affect responses. When LB pleasure does not materialize, need is distressing, but qualitatively different than punishment. When love feels betrayed, hate becomes a motivator; in the same way, disgust is a feeling of no longer wanting the same satisfier one wanted before. Frustration, a RB punishing affect, deals with pushing the boulder up the hill of social reality again and again.

I think of superego as Freud did: an ongoing evolution of the brain's affective responses to culture.

Freud's *The Unconscious* (1915) addresses a disbelief in unconscious thought: "Consciousness makes each of us only aware of his own states of mind ... without any special reflection we attribute to everyone else our own ... consciousness as well ... this identification is ... extended by the ego to other human beings, to animals..." (p. 169). "This ... inference ... applied to oneself in spite of internal opposition, does not ... lead to the disclosure of an unconscious; it leads logically to a second consciousness which is united in one's self with the consciousness one knows" (p. 170). I think Freud was trying to explain his split consciousness. From a dual hemisphere perspective, consciousness shifts between LB pleasure/unpleasure and RB affective pain/relief. The psychology of conditioning poses only pleasure and pain, treating unpleasure as in the same category as affective pain. Freud sometimes accepted that proposition.

Freud's guilt seemed to arise outside him. Unable to place guilt over Julius's death in himself, Monika, or the one-eyed doctor he projected it onto aboriginal man: "The content of the *Ucs* (unconscious) may be compared with an aboriginal population in the mind. If inherited mental formations exist in the human being—something analogous to instinct in animals—these constitute the nucleus of the *Ucs*. Later there is added ... what is discarded during childhood development as unserviceable ..." (p. 195).

In "The Neural Base of the Dynamic Unconscious," Heather Berlin has a different approach (2011). She explores how nonconscious cognition changes emotional attitudes without ever reaching consciousness. She considers awareness as a product of prefrontal linguistic reflection, a process that may not depend on consciously perceived stimuli. Indeed, sub-threshold perception can affect behavior. She cites subliminal images that whet the appetite. Freud knew the *Poetzl phenomenon*: sub threshold visual images presented to people to whom suggestion was made that they could dream the image did dream it unseen. Once the brain has built up its modes of habituated consciousness they operate together as networks even when some of the inputs are missing or when organic damage has destroyed some of the network nodes. For instance, the insular cortex can be accessed addictively by sub threshold stimuli. Berlin asks how these phenomena jibe with Freud's dynamic unconscious. One answer is that multiple stimuli stay at a preliminary level of distribution until attention networks include them. Moreover, particular stimuli enter many nodes with multimodal receptors without being included in consciousness or their effects in any one way. Just as visual stimuli are distributed to many multimodal areas of the brain—not only primary organizers of vision—nodes in various identity-related networks add to their activation level outside of consciousness.

The hippocampus can sort stimuli without evoking declarative consciousness or even emotional episodic consciousness. Even after hippocampal destruction, when new long-term memory is impossible, a person maintains a sense of functional identity in areas of cortex involved in earlier identity-related choices about what to do or how to access other people's opinions. We have temporal areas of medial cortex involved in how past emotional choices worked out; and parietal areas of medial cortex involved in how past choices of behavior worked out. These areas can sustain prefrontal practical thinking without using new sources of data. These networks were used in the development of a theory of mind, which the prefrontal cortex uses in our social interactions. Thus our sense of identity—what it means to be human—can become conscious from the stimulation coming from these medial networks. Ego function networks such as these have a general effect on identity awareness—and they can generate imaginary scenes.

Freud put practical thinking, fantasy thinking, and traumatic thinking into a dynamic system of mental, cultural, and biological influences. He proposed that guilt began as anxiety about abrogating archaic cultural rules; guilt was selfish individuality when the community was the symbiotic source of identity. Narcissism was the enemy of culture. In 1926 in *Inhibitions, Symptoms and Anxiety* he concluded that anxiety was not just a signal of maintaining practical thought by defending against irrational wishes, but a forewarning of trauma. In addenda B, he posed birth trauma as prototypical helplessness against reality. In Addenda C, *Anxiety, Pain and Mourning* (p. 169) he discussed infant stranger anxiety as a precursor of *peek-a-boo* games, glimpsing the parents repeatedly to master anxiety.

BRAIN HUBS AS EGO FUNCTIONS

Interactive networks may form brain hubs equivalent to ego functions that organize the conscious survival repertoire. Given the plasticity of cortical processes, the more a hub is used, the more the neural activations and inhibitions facilitate future use. So, repetition of activations and inhibitions saves metabolic energy in the long run. Are the hemispheres two super hubs competing ambivalently with each other? When neuroscientists measure activations, they often find wide disparities in comparable areas of the two hemispheres.

The LB activates in pursuing appetitive needs; the RB during task management. But Raichle showed that the brain's overall energy use is negatively correlated between DMN and executive task management hubs. Its energy use is highest when the DMN is activated, behavior is inhibited, and consolidation gets underway. He says researchers failed to recognize this discrepancy since very slow waves generated when neuronal change is underway in DMN learning was treated as mere "noise," just as those who studied the human genome treated DNA non-gene producers as "junk." The brain is most active when the body is inactive. The brain's prediction failure and what to do about it requires daily changes in neuronal structure. Nielsen et al. (2013) show that when needs are unfulfilled, the RB switches on the DMN hub.

Stage change passages must be highly metabolically activated, because everything has to be relearned; while the past stage is relegated to identity past. The developing brain has so many neurons entering hubs and making new nodes that the developing hubs use a large amount of energy. The younger children are, the more REM sleep they get per 24 hours, which leads me to think that REM accompanies new emotional learning.

INDIVIDUATION, NARCISSISTIC INPUT, AND CULTURAL SUPEREGO DEVELOPMENT

Just as Darwin showed that animal communities survive by signaling food, water, or danger, as well as sexual or aggressive readiness to one another, culture plays the same role for humans. Identifications lead a person's cultural development. Freud showed that cultural narcissism expands because everyone becomes their own fantasy hero or heroine of culture. Freud viewed moral affects—shame and guilt—as the cultural inheritance that develops in stages. Preoedipal love and aggression form identifications with each parent as family culture. In latency, wider cultural identifications give rise to morality: "The Super-ego … originated from the experiences that led to Totemism … the id … harbored residues of the existences of countless egos, and when the ego forms its super-ego out of the id, it may … be bringing them to resurrection" (1923, p. 38). Freud was subjectively obsessed with revenants: imaginary forebears, muses, and ideal imagoes, the dead who were transformed into objects of cultural identification.

CULTURAL DISSONANCE PROMOTES REGRESSION

For Freud, identifications with the dead were a major social organizer contributing to the illusions of religion. I want here to consider not just the cognitive and emotional brain, but the cultural brain. Emotional conflicts are triggered by dissonance between culture's moral values and practices, and a person's family beliefs and mythology. We can see this conflict working in a religious divide between more secular and more fundamental sects. Durkheim wrote that social dissonance results from the loss of shared religion that had protected individuals from over-reliance on the nuclear family (1897 [1951]). Cut off from shared cultural morality and customs, many people develop *anomie—Who am I?* Some cultures avoid dissonance by spreading responsibility for socializing children into the general community. Israel extended families into Kibbutzim. But nations can destroy allegiances and create economic conflicts among its tribal groups. A prevailing culture cannot run the brain efficiently when the individual feels it is unreliable and foreign.

Language mediates between culture and the brain. Auditory signals activate multimodal networks throughout the neocortex, as well as centers for language reception, speech formation, and thought, put together from both hemispheres. Human language extended animal language through an evolved capacity to speak gestural consonants using lips, tongue, laryngeal, and glossopharyngeal muscles, combined with the prosody of musical affect punctuated by conscious breathing. But language also has signifiers, cultural meanings that can vary like idioms. It is much easier to lie with language than to falsify visual impressions of episodic experience. Freud knew that language represents experience within its declarative realm of what can constitute false memory. Dissonance between lived experience and falsified experience adds to culture's dynamic unconscious.

Individual identity is formulated in words and iconic images. Like a semicolon parting two forms of consciousness, the LB is unconscious of threat requiring defenses and suspension of pleasurable activities; the RB is dynamically unconscious of the LB's pleasure needs. Since the sense of self includes both vantages divided by the hemispheric subjective/objective split—two forms of consciousness side by side alternate; self remains a slippery experience. Like the rhetoric of an impostor, ideas that the RB opposes may stay unvoiced. On the other hand, cultural power retains taboos, injunctions that can turn a superego against one's social self, a mind divided against itself. Perhaps it is only in DMN fantasy that the two sides of self engage in a creative experiment of becoming more like someone else.

DERIVATIVES OF FREUD'S CULTURAL UNCONSCIOUS

For Freud, ambivalence was an innate quality of both the brain and the mind. The archaic group was bound not just by the pleasure of dance, music, sex, and ceremony, but also by taboo and safety. Freud realized that every individual resents his own cultural persona: "… the intruder was nothing but my own reflection in the looking-glass on the open door. I … recollect that I thoroughly disliked his appearance. … Is it not possible, though, that our dislike of them was a vestigial trace of the archaic reaction which feels the 'double' to be something uncanny?" (*The Uncanny*, 1919, p. 248).

The double is one's own RB self objectified as other. In *The Uncanny* Freud analyzed primal ambivalence as the projection of spirits of the dead kept alive in the unconscious as part of the ambivalent identification with the other. Freud explored *The Antithetical Meaning of Primal Words* (1910) in archaic languages, like innate ambivalence between the sacred and the profane, which shows that superego originates in id. Relinquishing revenge for the sake of cultural cohesion gives a moral sense of safety from anxiety.

Antithetical primal words show innate cultural ambivalence: *heimlich*, familiar or homey, also means *unheimlich*: strange. Like Freud in that railroad car, the image of an *unheimlich* person enters our space—our self in a mirror. We have met the enemy and he/she is us. Denying ambivalence toward our own social self, we all want to believe it is others who are aggressive to us. Freud's turned Julius into an archaic figure of self-transference who accompanied him on his psychoanalytic Odyssey.

Freud's preconscious system manifests how pleasure discharge is latent before it can happen. There is time for the RB to inhibit an impulse that turns into intention; signal anxiety intervenes. RB damage can cause impulsivity. The time lapse between RB perception and LB action, between raw libido and aggression—like an impulse to rape for instance—is the neural space for binding the mental system by the superego.

HOW AUTONOMIC EVOLUTION ENHANCED MENTAL PROCESS

Supporting mental-system expansion, culture co-evolved with the brain's autonomic, parasympathetic, and sympathetic systems. The parasympathetic system developed to prioritize satisfactions required for metabolic replacement as resources are used up, and to ensure sexual reproduction; while the sympathetic system evolved to assess and respond to safety issues and to communicate effectively with others. The autonomic system mediates mental responses when the limbic system signals go ahead, wait, or stop. Dopamine neuromodulation drives LB pathways to appetitive action. Basal ganglia centers are dopaminergic in their LB regulation of the pleasure inherent in behavior (Molochnikov and Cohen, 2014); while RB norepinephrine neuromodulation mediates signals for cautious, even vigilant attention to the environment before acting. The right ACC is anti-correlated with left OFC pleasure activation (Craigmyle, 2013).

Bud Craig, an expert in functional brain anatomy, reports that the limbic AIC that underlies the lateral PFC integrates mental progress toward goals (2005). It unifies social, cognitive, and emotional consciousness, responding to difficulties in carrying out need functions. I think of the AIC as a limbic synthesizer of practical thinking. Craig shows it mediates between prefrontal consciousness and the autonomic system: The left AIC promotes feeding and routine behaviors in all vertebrates, the right AIC triggers negative effect, avoidance, arousal, and avoiding predators. Its mediation is fortified by a high volume of Von Economo neurons, which facilitate mental functions.

Mammals are distinguished from reptiles by a shift to social communication of affect and by the capacity to delay responses to the social and physical environment during a period of assessment. They increased their capacity for thought with increased neocortex, as their autonomic nervous system evolved prolonged attention, motion, emotion, and communication requiring enhanced oxygen intake. The renowned neurophysiologist, Stephen Porges, explained how our sympathetic and parasympathetic systems evolved in tandem (1993). The parasympathetic vagal nerve evolved a new effector center in the medulla that maintained cardiac and respiratory function during novelty through its effect on the sinoatrial node, the endogenous pacemaker of the heart. That "smart" vagal input of the NA (nucleus ambiguous), brings the heart and respiration into phasic cadences that maximize oxygen availability for higher cortical work: *"Primary emotions are related to autonomic functions* ... related to survival, they must be integrated into cardiopulmonary regulation ... primary emotions have a right hemisphere bias ... with the regulatory ... medullary structures controlling visceral function" (p. 314).

The smart NA is called into play when we are challenged to explore the unknown. The mammalian NA system coordinates a RB shift to an enhanced capacity to think and act flexibly when novel situations arise. In challenges, a reptile's parasympathetic system simply slows the heart to preserve metabolic resources. That tonic mechanism enters into the human reorientation reflex. But in situations of extended surprise, the reptilian tonic slowing system can be too prolonged for humans. Porges links infant death syndrome to vagal sinoatrial slowing due to immature NA development. The smart NA enables us to use feelings that have signal value, providing conscious phasic control to the innervation of the larynx, trachea, and esophagus, and the facial muscles of expression. Porges makes a strong case for the smart vagus as a major source of RB affect expression:

> The functional dominance of the right side of the brain in regulating autonomic functions and emotion may have implications for the specialization of motor and language dominance on the left side of the brain. The right-sided responsibilities of regulating homeostasis and modulating physiological state in response to both internal (i.e. visceral) and external (i.e. environmental) feedback may contribute to the development of motor and language functions on the left side of the brain. (1993, p. 313)

Human social evolution and the use of mental flexibility in assessing reality require high metabolic energy resources. If the RB controls decisions about life and death behavior it is useful to have the LB using language and logic to help process the complex decisions.

Dramatizing the innate calming effect of the RB smart vagus, Porges describes the effect of lullabies on the infant, even the premature infant (2011). He writes: "Vocal music duplicates the effects of vocal prosody ... neural mechanisms that regulate the entire social engagement system with ... changes in facial affect and autonomic state. Basically we start to look and feel better when we listen to melodies" (p. 210). Archaic human communication was vocalized in song. Our ancient forbears were soothed when medicine men, shamans, and later bards like Homer sang of the issues surrounding survival and death. The RB's control of prosody and affect tone in speech helps spoken communication maintain order in the brain and in the social community.

Human evolution developed resources to mull trial action and social response. Instincts triggered by the central amygdala are kept on hold during novelty evaluations. Porges distinguishes biologically based emotions like fear and sadness from culturally determined emotional responses like doubt, guilt, and shame. Those feelings trigger reflection on one's social identity and how to respond to other people. Cultural systems evolved prolonged group assessment of environmental challenges, as the smart NA inhibits instinctual responses. Individuals with an inadequate smart vagus may revert more easily to instinctualized fight or flight. The NA links to the amygdala, the PAG, the ACC, the hypothalamus, and the hippocampus, all sources of emergency trauma response (Porges, 1993). Measuring heart and respiratory rhythm during challenges to children's capacity to respond to emotional signals, Beauchaine, Gatzke-Kopp, and Mead, found those with conduct disorder or ADHD had less vagal flexibility than a control group (2007). The parasympathetic polyvagal system inhibits behavior while subjects mull responses to social novelty (Beissner and Meissner, 2013).

THE EXPLOITATION-EXPLORATION MODEL

Things get interesting when predictions fail. The evolved sympathetic system brings us into a sense of adventure, which the smart vagus can prolong. If *exploitation* is LB behavior, using dopaminergic pleasure centers to take the best of what is present in the environment; *exploration* brings new possibilities, a bound emergency instinct that facilitates seeking. Gary Aston Jones and Jonathon D. Cohen's exploitation-exploration model (2005) does for our understanding of sympathetic brain function, what Porges did for parasympathetic function in mammals. Just as our parasympathetic system evolved to deal with uncertainty, our sympathetic system also has an ancient tonic mode and an evolved phasic (conscious) mode enhancing attention to novelty.

Porges showed that respiratory rhythm during attention measured parasympathetic vagal stimulation. Jones and Cohen use eye-pupil dimension to measure tonic LC (locus ceruleus) sympathetic increase (widening). Novelty induces orientation, a sympathetic vigilance response that widens the pupils, followed by an exploratory phasic response with mild narrowing, during a search for what went wrong. When something new is apprehended, mild widening occurs. The right ACC links to the locus coeruleus, its source of norepinephrine (NE), which determines the brain's sympathetic tone.

While the LB dopamine system adds pleasure in each sub-goal task success, the RB's NE system uses painful signal affect to alter failing tasks:

> ... the LC phasic response can be thought of as an attentional filter that selects for the occurrence (i.e. timing) of task-relevant events and facilitated responses to these events. But the tonic response increases with mistakes and difficulty, allowing more possibilities of new choices. ... (Jones and Cohen, 2005, p. 416)

Eldar, Cohen, and Niv explored NE effects on social task enhancement (2013). During a game requiring social collaboration to get the best reward for both participants who had to choose LB semantic or RB visual cues, the subjects' pupils would widen when they made temporary gains using visual cues. Such a sympathetic system reward is relief: "Oh, I got it." The ACC plays a pivotal role between the parasympathetic vagus and the sympathetic LC, in making action choices. Reviewing how the autonomic system sets mental priorities, Craigmyle notes that ACC activation is responsible for the P 300 (300 milliseconds) EEG response to an object (2013): "Is that what I am looking for?"

Controlling off/on switch modes, the right ACC determines how much NE will be allocated to salience, attention, executive functions, exploration, learning, long-term-memory, creative fantasy, and stress response. Craigmyle sees ACC as the fulcrum of RB mental functions; because it participates in *salience* hubs, *executive* hubs, *attention* hubs, as well as an *affect* hub that links to the NA, LC, amygdala, and PAG. Thus, the ACC is a major limbic link to medial emotional functions, while the lateral positioned AIC (anterior insular cortex) is a major link to cognitive organization; together, the ACC and the AIC insure the brain's practical work.

Cognitive familiarity is a sign of potential satisfaction; while unfamiliarity may be a sign of a lack of needed salience. Bud Craig (2005) and Anthony Damasio (2003) link addictive pleasure to left AIC and affect responses to right AIC salience detection. Pleasure, frustration, and salience detection require bilateral collaboration of the hubs involved in the presentation of consciousness to its PFC regulation. The AIC and the ACC are the limbic cortical areas that work together to mediate feelings processed by the VMPFC, along with the orbital frontal cortex (OFC). These three limbic anterior cortices form and affect social conditioning.

Cultural learning can be communicated by people who are more expert in reflecting on the processes of their own relationship to satisfaction and suffering. For instance, studying awareness in mind-wandering, Craigmyle found that experienced meditators alternate executive and DMN functions. They stay aware of ongoing events and the potential impact of their social attitude. Similarly, if you ask a sensitive psychoanalyst what they were thinking when listening to a patient, he or she mentions the impact of their own inner thoughts on the patient. Perhaps that corresponds to regression in service of the ego, in which people keep their capacity for observation intact throughout a stay in the DMN.

Much is written about empathy as mental communication that locks us into other's brain process. Putting mirror neurons aside as conveyors of LB communication about the readiness of satisfiers, let's focus on RB empathy, which brings a person into feeling another's painful affect. If we see kin in physical or emotional pain, we feel it and process it in the right ACC. I saw many drug addicts who left their dear ones and lived on the street, because they couldn't bear that their addiction would ruin the life of their loved ones. I think of songs that stay with one all day with a RB affect of suffering or relief.

GRADATIONS OF TRAUMA

Considering the vicissitudes of aggression, we find LB aggression exploits others for narcissistic advantage and pleasure, and RB aggression responds to an adversarial attacker who threatens survival by producing imminent trauma through attack. The right ACC triggers a series of steps in response to unexpected situations and to potential trauma, steps which recapitulate the autonomic evolution that prepared the mental system to follow a necessary process. As we have been seeing, the right ACC responds to novelty with bound aggression, enhanced attention, and vigilant exploration. But in a survival emergency, practical and fantasy both suspend as the RB's ACC and amygdala trigger sequenced instinctual responses leading to fight, flight, or paralysis.

The right sympathetic system triggers a graded array of responses to anxiety signals, affective surprise, vigilance, fear, and stress, before emergency response. In prolonged novelty the right ACC links into posterior cingulate cortex [PCC], which evaluates challenging cultural, cognitive, and emotional dissonance. Fornita et al. explored links between the DMN and the right PCC's link to the hippocampus (2012): "… the right PCC seems to act as a critical information processing bottleneck, representing a major hub of the transitional DMN module, while also retaining high functional integration with other regions" (p. 12791). The dorsal PCC filters cognitive mental data used in task management; the ventral PCC, a major DMN node, assesses prevailing emotional data. The salience networks and the DMN's social memory can rapidly alternate their activation.

Since executive functions do not have their own direct links to episodic memory, we go in and out of DMN meditation to be in touch with direct recollection. I am aware that my long-term memory fantasy time may be as brief as a few seconds, or go on for timeless minutes. But the task mode and the DMN capacity for retrieving episodic memory coexist during normal transition. Thus, rapid alternations may allow episodic memory to be placed in short-term memory to become available to problem solving. Fornita et al. review data showing that separate states of cognizing reality and meditating in the DMN may disappear in psychosis and traumatic brain injury (2012).

MYTH: THE FANTASY OF CULTURE

Freud formulated that the survival of the individual and the species required cooperative kinship and tribal harmony. Like other mammals that evolved gestures, expressions, places to live, and their own social communication, humans build social structure. But in humans, identification with kin, especially with parental figures and with chiefs who watch over the whole clan, leads to identification with the dead. Human neocortex can generalize the lessons of forbears' into a template of personal identity. Shamanistic lore, for instance, worshipping totems from generation to generation, gave them a survival advantage—those who pray together stay together near the best water holes.

Ancient writing, such as marks on stone and clay, began with inventories of crops and goods; and in ancient Egypt showed work accounts and the social role of the worker. Yet, as with the 5000-year-old Epic of Gilgamesh, writing was also used to glorify dead leaders as heroes who deserved to live after death. That Epic was rediscovered and revised by cultural historians 3000 years ago. Perhaps shamans also practiced the art of tattoos to mark the survival role of particular men and women of the tribe. Virtual beings, the dead live on as phantoms, phantasies of eternal survival.

Phantasies that bring the living and dead together persist in our era. I had a patient who was a male prostitute during the AIDS epidemic. One of his clients served a lavish feast for just the two of them in a Pacific Coast mansion as the sun was setting over the ocean. My patient had to stay outside until nightfall, then return to find his *master*—somewhere in the dark of the mansion. After a long search he found him in a coffin on the steps of a second landing. It was there that the penetration happened. Afterward, he was never to return. Yielding self in anal orgasm (as other patients have told me) may have been the master's way of breaking free from executive tedium. Was his fantasy of joining the dead his rejuvenation? Was the feast a last sacrifice?

Ego functions are controlled by executive identity functions shaped by culture. The way I see it, culture is absorbed through bilateral superego functions: left-brain shame and right-brain guilt. Freud reasoned that both affects begin as cultural anxiety. Shame is a feeling of failing to work together for tribal survival, which in later stages became failure to live up to cultural ideals; while guilt indicates breaking the rules of conduct toward others, which results in a loss of status requiring punishment. To be shunned by the tribe was worse than death. Freud concluded that internalized anxiety developed in human prehistory as a signal of such cultural breaches.

WHERE IN MY ID, EGO, OR SUPEREGO IS MY SELF; AND WHERE ARE YOU?

Trauma-induced survival fear results from a person's loss of capacity to deal with the physical duress of body damage, or from the loss of others who are felt to be critical to one's emotional survival. Prolonged acute stress depletes executive ego functions as the RB begins a cascade of phenotypical changes in its ACC, amygdala, and VMPFC, which collectively induce the loss of one's capacity to record fresh experience that might block trauma. Biological processes engaging the midbrain during acute stress alters autonomic control, deregulating mental process. Akirav and Maroun describe self-depletion in the aftermath of rage or a panic attack (2007).

Describing the relationship among ego, id, and superego, Freud looked at social evolution in the context of repeated experiences of cultural regression and development. As he conceived it, people inherit collective cultural experiences that determined the stages of childhood by recapitulating past stages of cultural development. He often uses the metaphor of the mind being layered like the ruins of civilizations, such as he saw in Rome, where one could reconstruct layers of cultural development by digging from the top-down. Within that fantasy he embraced Carl Jung's concept of an inherited collective unconscious. His kinship with Jung so permeated his cultural unconscious that he fainted on the boat to America: feeling that he and

Jung—like a younger brother—shared the same dreams. He wished Jung to be his standard-bearer for the future of psychoanalysis.

Yet, Freud always dismissed his necessary companions, who functioned as muses for him, replacing them with his own logic. In his lifelong self-analysis he became aware of the rational/irrational dimension of every concept—which he ascribed to the brain's inherent dualities. He saw this in the innate opposition of primal words like *hiemlich/un-hemilich* (familiar/unfamiliar), which he ascribed to mental projection of the brain's developmental processes onto culture. Libido/aggression, love/hate, wish/defense, initiate/imitate, and verbalize/image, activate separate hemispheres: LB/RB, respectively.

Since self is inexorably tied to others, I consider identity functions from that perspective, and here I want to offer a few ideas you will encounter throughout the rest of the book. First, the ego is a prefrontal, organized mental system that compares its identity states to body signals of affect and anxiety that gauge progress toward goals. Second, the id is a set of neural centers with diverse hormonal, neuromodulator, and genetic phenotypical effects on their specified cortical networks. Third, the superego represents the cultural world's impact on the executive identity ego. Social anxiety evolved into affects of guilt and shame that signaled moral failings. Freud tried to discover how cultural development led to equivalent brain processes.

In *The Ego and the Id* (1923), he treats superego as an archaic heritage producing moral values, despite its origin in the id. He ends a section on the superego and the ego: "The struggle which once raged in the deepest strata of the mind, and was not brought to an end by rapid sublimation and identification, is now continued in a higher region, like the Battle of the Huns in Kaulbach's painting" (p. 39).

After World War I, Freud feared that *progressive* culture was threatened with the aggressive culture of the archaic past, carried out by modern-day Attilas. He could not resolve the nature of the unconscious, because he tried to understand it in the physical structure of the brain rather than as a response to culture. The unconscious is the domain of cultural structure, not physical structure. He knew this but he didn't know it. Chapter 5 examines the origin of cultural narcissism in the form of Gilgamesh, Freud's alter-ego when he was in the midst of his own heroic phantasies of creating psychoanalytic culture.

Chapter 5

From Gilgamesh

The Oldest Culture We Know

The historian Norman Davies suspects a unifying mythology in the cultural chain of human history. Beyond instinct, preserving the mind of the leader as *geist* was a form of group memory preserved in songs by poets so that in time they became the possessions of individual memory. As I see it, Homeric epics preserved libidinal stories of Odysseus's wanderings in the Odyssey, and aggressive ones in the Iliad. *Gilgamesh* is far older, sung and told from 2750 BCE, then reproduced on tablets and eventually standardized with added pieces from other epics around 1200 BCE. The concern for passing along life experience to future generations shows reflection on the relation between individual life and the life of the familiar group, of culture itself—the two sides of human identity.

Here is how Davies puts it:

> The opening invocation of *Gilgamesh* resembles the opening lines of the *Odyssey* both in tone and sentiment: ... Stronger still is the case that can be made for the influence of *Gilgamesh* on the *Iliad.* Both epics turn on a twist of the plot which occurs with the death of one of two inseparable friends. Gilgamesh mourns for Enkidu as Achilles mourns for Patroclus. ... If this supposition is correct ... the Homeric epics ... also span the gap between the conventional western literary canon and the far more ancient writings of non-European literature. (Norman Davies, 1996, pp. 114–115)

Chapter 5 pursues the origins of narcissism in the formation of individual contributions to culture. The Bible, but earlier than that, *the Epic of Gilgamesh*, presents these origins as moral lessons about leaders' responsibilities to culture. *Gilgamesh* was on Freud's mind as he wrote *Totem and Taboo*, his seminal work on the origins of culture. He was contending with

Jung for the mastery of cultural origins and for being the leader of the psychoanalytic movement as it included culture in its purview. Whose mind would live on as the leader of psychoanalysis? Gilgamesh deals with the eventuality of death, dramatized by his attempt to find eternal life and to take his place among the gods. That wish for permanent survival of the individual's narcissism taking root in eternal culture distinguishes individual death from the death of culture. The memory of kin and past leaders can live as a vital part of civilization itself. We will see how much Freud's fantasy of legacy and constructing a cultural understanding for psychoanalysis was embedded in his DMN thinking as he mulled over the myth of Gilgamesh.

How Culture Runs the Brain views our anterior cortex as the organizer of human culture, absorbing cultural communication and internalizing its precepts within the sense of each individual's identity. A social mind of familiar self, familiar others, and strangers who may help or threaten one's group has a virtual quality—like a concept of god in the mind as culture. The LB individual self looks to the future as an extension of motivated effort and goals, while the RB objectified self, as part of culture, feels eternal like all who went before, which is one reason people sacrifice for the good of group survival. Fantasy and myth are imbedded in the wish to keep the dead in your heart. The cultural leader is an individual; while the shaman presides over the collective phantasy of eternal life and its ceremonial celebration. These roles respectively represent practical thinking and fantasy thinking.

RECONSTRUCTING THE DEAD

Primitive man cannot think of death except as the sudden departure from the dying person from that which originally brought life. Nevertheless, there is evidently bound up with this conception the further idea that powers of life are still resident in the body; the latter remain firmly associated in the mind of primitive man with the impression of life (Wundt, Wilhelm, *Elements of Folk Psychology*, 1916, p. 82).

Freud absorbed Wundt's ideas. Individuated identity can lead toward feeling a loss of group membership. The anxiety this evokes triggers a state of mourning for culture, or a soothing fantasy to restore a sense of social identity—like Freud's making new culture to overcome his losses as an infant. We build civilization trying to compensate for death with religious geist. Social spirit infuses the collective group with a generalized unity of souls and it inspires institutions of culture. The archaic feeling of avoiding death was tied to identification with the individual totem, which bound people with nature. Eating the dead totem gave rise to sacrifice. Having a mind of your own can feel dead wrong.

Totemic hordes developed shamans who maintained myths, gods, and culture of the dead, through customs and ceremonies that provided rules, guidance, and hope for the living. The Gilgamesh Epic tells the mythic development of culture, from hunter-gatherer origins in the forests, to pastures and mountains where sedentary cultures could survive. Uruk in Mesopotamia was an early fortress of sedentary culture. Animal gods were part of the mythic past of Uruk culture, being replaced by humanized gods. The *Epic of Gilgamesh* narrates the passage of culture from animism toward trade and humanized gods.

The Epic warns against too much individuated narcissism in the leader that may disturb the cultural functions of holding on to the past. The Uruk culture used cuneiform to record its history for hundreds of years before the hero-king's reign some 3250 years BCE. As hero, Gilgamesh portrayed a man modern for his times who worked to help his culture progress. Yet he was too solitary in his decision making. He faced turmoil from his Uruk people due to external attacks and his despotic control. Consequently, he made an alliance with a pastoralist nomad leader. Songs of praise express how these two heroic leaders merged their cultural strengths through love, trust, and shared gods. Gilgamesh, meaning *the old becomes young*, identifies with Enkidu, an image of perpetual youth who gives up his life so that Gilgamesh could keep Uruk progressing—like Freud's relationship with Jung, who was a revenant of Julius.

Enkidu emerges from the hunter-gatherer culture of the forest, onto the pasture land where he is totemically identified with animals that occupy the interface between forests and pasture land. Not knowing human parents, wild asses and mules are personal gods who carry him from place to place. Gilgamesh, King of Uruk, is a mighty warrior who rules by strength. To show his culture's vitality he wants to construct great cedar doors on the walls surrounding Uruk. To carry out this feat he needs Enkidu, who knows how to fight Humbaba, the looming presence in the cedar forest who inexorably kills strangers seeking to penetrate his archaic domain.

TABLETS: CULTURAL LIFE THEN AND NOW: FREUDIAN CORRESPONDENCES

Tablet 1 of *Gilgamesh* begins: "He who has seen everything, I will make known to the lands." The Epic presents the vital core of past culture to leaders of Uruk's future who can read and use history to help culture progress. Portrayed as the heroic son of the goddess of cattle and the king of Uruk, Gilgamesh improves the largest City-state known, founded on the Euphrates some 500 years before his birth. He upgrades the wall around Uruk with copper; opens mountain passes; digs wells from the mountain side to irrigate crops; and expands the Temple of Ishtar to bring fertility.

Death and aging are unacceptable to Gilgamesh. His narcissism is so exclusive he cannot share his life with a wife to ensure the line of Kings. His narcissistic fantasy is to live forever, embodied in culture, bringing the past into the future, bringing the old into youth and youth to the old. Into the breach, his goddess mother creates a mighty match for him—a being that has a past history. Enkidu will be like a wife, a double from past culture. Leaders emerging from the primal horde with LB qualities of primal narcissism can join RB totemic gods and goddesses of the animal kingdom. Right-brain gods and goddesses are miracles: religious phantasies everyone can believe.

Shaggy with hair and jostling at the water hole with other animals, Enkidu was mighty as the totemic beasts of the wilderness. Yet, to partner Gilgamesh, he had to acculturate to Uruk. The Temple of Ishtar opened Uruk to other cultures through trade and provided temple harlots. Ishtar, who had her own Epic written in the eighth century BC, knew how to rein a man's savage instincts and encompass his passion, by satisfying him until he was tamed. Shamat, her harlot, continuously satisfied Enkidu's lust for six days and seven nights until he was sated—and his animal kin shied away. In the Epic, forgoing sleep for six days and seven nights marks a capacity for eternal life.

DREAMS OF HOMOSEXUAL ORIGINS OF CULTURE

Gilgamesh dreams that a meteorite of the sky god brought an object so strong that he cannot do anything but lay it at his mother's feet. The Uruks throng about it, kissing its feet like a baby's. His mother says the dream means a mighty new friend, close as a wife, will save him—and Uruk. In the second dream the Uruks throng around a magnificent axe that he lays at his mother's feet. He says he loved it like a wife. She interprets it to be a man who will be his wife, a mighty competitor, and comrade. Since the Sumerian word for axe and wife are the same, composers of the Epic may be playing with puns—as dream formation often does. But in my view, the myth is of a younger son who will replace the aggression of the older son and rule with a mother's kindness. After Shamat tells the dreams to Enkidu, they make love again. Telling Enkidu how strong Gilgamesh is, she challenges him to be his equal; teaching him to dress and speak Uruk: how to grow up.

Enkidu confronts Gilgamesh by blocking his entrance to the chamber where he deflowers maidens about to marry. They fight to a standstill of embracing each other. Foretelling this, Gilgamesh's goddess mother had defied Ishtar. She knew that recoiling from female control together; Gilgamesh and Enkidu would promote their alliance, just as Freud would later know that male homosexual bonding to promote culture mitigates the competition that leads to raw aggression. Freud thought that youngest sons reflect their maternal bond through their cultural sensitivity in dealings with other men. He carried the spiritual identity of Julius throughout his life; and like Gilgamesh, he always idealized his almost mythic cultural forbears.

To promote his mythos, Gilgamesh plans to build a huge cedar door for the ages that can open Uruk to other cultures or close it for safety. In order to cut cedars and float them down the Euphrates to Uruk, he wants Enkidu to help him kill the forest's demigod, Humbaba. Enkidu says "Humbaba's roar is a Flood, his mouth is Fire, and his breath is Death!" After huge travel each day, Gilgamesh wakes with anxiety dreams each night:

Calling out helplessly to Enkidu he asks him why he felt trembling and alone facing an omnipotent god. He tells Enkidu he dreamed of the mountain falling on him. Like their mother, Enkidu can interpret dreams, saying it means Gilgamesh will bring Humbaba down and throw his dead body in the wasteland. Just then Humbaba's roar terrorizes them. Gilgamesh overcomes his anxiety, telling Enkidu: "Let your voice bellow forth like the kettledrum, let the stiffness in your arms depart, let the paralysis in your legs go away."

Humbaba makes fun of them with the aggressive imagery of forest people whose totem gods were vultures and eagles. He uses mocking humor to fight for his forest's life by asking how an idiot Gilgamesh and a moron Enkidu, who no more knows his father that a fish, can challenge him when he himself preserved Enkidu's early life. Despite pleading to become Gilgamesh's servant they kill him; as he curses Enkidu to die. Gilgamesh decapitates Humbaba's head as trophy, and cuts down his cedars for legacy. Because he killed Humbaba for glory, the great god, Enlil, will take Gilgamesh's life in return.

For his triumph, Ishtar wants to make Gilgamesh her husband. Knowing she will castrate his narcissism, he refuses. Incensed because she gave the people of Uruk their fertility, she calls upon the totem of Uruk, the bull of heaven, to kill the men of Uruk. But Enkidu and Gilgamesh kill the totem bull, throwing its hindquarters in her face. Gilgamesh mounts its head and horns on his palace and he gets the men of Uruk to feast on its body—just like Freud's reconstruction of the archaic brothers' oedipal feast. Who will be the mother's favorite, the first born son as Freud imagined, or the youngest son?

A MELANCHOLY ALL CAN SHARE

Enkidu dreams that the council of gods decides Gilgamesh must die, but he puts himself in his friend's place. For attacking the goddess of fertility out of love for Gilgamesh, Enkidu must die. Anzu, the storm god seizes him with paws of a lion and talons of an eagle: Gilgamesh is afraid to help. Mourning

Enkidu, Gilgamesh enters the wilderness of melancholy, tearing off his civilized clothing. To save Uruk he must find Utnapishtim, a Noah figure, who, with his family and all the animals, survived the great flood and repopulated the earth. Gilgamesh wants to know how he too can defeat death.

In his quest he dreams of the technological artifacts of culture's power: axe, sword, and arrows, to kill the wild animals that would devour him. Wearing their skins, he identifies with Enkidu, but he lacks Enkidu's empathy with totems. In the narcissism of failing morality, he will not acknowledge that all of god's creatures are his kin. When he comes to the tavern keeper, Siduri, who lives on the shore of immortality, he appears as a crazed murderer. But he elicits her sympathy with his pitiful tale of mourning, and she tells him how to find the ferryman Urshinabi.

Gilgamesh cuts huge punting poles from the forest to avoid the death waters in crossing over to Utnapishtim. Utnapishtim reveals that he received Enlil's blessing for eternal life by staying awake, like a god, for the seven nights and six days of the deluge, while caring for his humans and animals. Given the same opportunity, but consumed by mourning (as Freud often was), Gilgamesh falls asleep. Sorry for him, Utnapishtim and his wife, tell him of the plant of eternal youth at the bottom of the sea: named "The Old Man Becomes a Young Man," a drug of eternal youth (like Freud's illusions about cocaine). While Gilgamesh bathes to restore his cultural self, a snake eats the plant and sheds *its* skin.

FREUD AS GILGAMESH

Freud wished to be a hero like Gilgamesh, infusing his life force into cultural progress. In 1911/1912, while working on *Totem and Taboo*, his relationship with Carl Jung reached a crisis. For several years Jung was Freud's heir-apparent, the Christian who could bring psychoanalysis into the whole collegial world of progressive culture: Jung already worked with Eugen Bleuler, Karl Abraham, Ludwig Binswanger, Adolph Meyer, Herman Rorschach, and Abraham Brill at the prestigious Burghölzli psychiatric hospital. Freud and Jung enacted a mutual dance of professional seduction for several years, until their interests merged in reconstructing origins of religion in culture. Whose theory would prevail?

Freud had been trying to convince Jung that sexuality is the core of the libidinal instinct, telling him it fueled the progressive developmental stages of oral, anal, urethral, and mature sexual organs, leaving fantasies behind that determined personality organization. But Jung treated libido as a general category of human motivation including such diverse biological mechanisms as hunger, thirst, and even respiration. Perhaps like Panksepp and Solms, he saw motives as innate givens, with symbolic content that originated in the biology and culture of animal development. Jung had become Freud's irrational muse, who had to be overcome by logic. Like Fliess's calculation that male and female periods beginning at one's birth determines one's health crises and the total period of one's life, Jung believed that the confluence of stars at one's birth determined one's cultural place. As Freud became disaffected with Jung as his doppelganger, he turned toward Ferenczi and Otto Rank as his muses.

Increasingly ambivalent about visiting Jung in Switzerland later in the summer of 1911, Freud took a trip to Sicily with Ferenczi instead. Discussing whether he should train an heir to replace him as the leader of the psychoanalytic movement, Freud blocked out the name *Jung* as an association to the name of a place in Sicily. He made this into an anecdote added to *The Psychopathology of Every-Day Life* (1901), confiding it in a letter to Jung that invoked the Gilgamesh Epic:

> … you will find an example of name-forgetting that occurred in a conversation between Ferenczi and me; to you, I am sure, the explanation will reveal an additional meaning that would not be apparent to others. It is the old mythological motif: The old god wants to be sacrificed and rise again in the new one. (McGuire, 1974, [*The Freud/Jung Letters*, Letter # 266, July 21, 1911], p. 436)

Freud substituted *Castrogiovanni* for Castel-vetrano—*giovanni* sounds like giovane, "young," and *vetrano* like veteran, "old." Jung might think that he will be the new god. Like Gilgamesh Freud wishes his own name could mean the old becomes young.

Freud goes to a lot of trouble to cover his tracks, but also to reveal them. He had not yet told Jung that he will not be his heir. Letter #266 ends with Freud's telling Jung in a joke that hides the truth that he has decided to endow a votive tablet like the Gilgamesh Epic: to be free of his ailments and to become young again. It seems that unbinding aggression for his pending break with Jung makes him sick—like his death wishes toward Fliess. Freud tells people what they do not want to hear in indirect way, which they can hear only if they are real men.

Racing to get to the bottom of human cultural origins, Freud was writing *Totem and Taboo*, while Jung was writing *Transformations and Symbols of the Libido* (1956 [1967]). He tells Jung in a letter dated August 21, 1911 that he "unearthed strange and uncanny things and will almost feel obliged *not* to discuss them with you" (Letter #268, p. 438). Freud visited Jung and his wife Emma on September 16, 1911 for four days, during which they discussed the *Epic of Gilgamesh* in a way that dramatized the end of their collaboration.

Jung had served his purpose of muse to bring Freud down to an unconscious realm where he could explore the cultural unconscious. Freud cut Jung off Jung just as he had killed off his doppelganger Julius.

In his letter to Freud after the visit, Jung wrote: "For our interpretation of the Utnapishtim episode in Gilgamesh I have found some rather weird parallels which shed light on Utnapishtim's gnomic utterances" (Letter #271, October 4, 1911, pp. 444–445). Utnapishtim became a god, while Gilgamesh failed in his attempt—which foreshadows Jung's psychotic delusion of being a god. Freud's next letter tells of forcing the "Adler gang" out of membership in the psychoanalytic society, adding a rebuke to Jung, by writing that Dr. Spielrein (Jung's analysand and lover) showed up unexpectedly in Vienna, saying Freud didn't look as malicious as she imagined. Freud refused to have his persona diminished by Jung.

Freud's letter #274 to Jung on October 13, 1911 uses their conversations on Gilgamesh during their last meeting to make Jung a nonentity in Freud's future. In the letter Freud treats Jung as if he is already dead and gone, while he continues to use the motifs of literature and myth to reconstruct past cultural life. He compares himself to Gilgamesh and Jung to Enkidu in such a way that the older brother is the real man, like Don Quixote was to Sancho Panza (who was merely a paunch). He reconstructs the *Epic of Gilgamesh* to write that the weaker younger brother dies first—indeed he is nothing but an afterbirth, a placenta that gave its libidinal energy over to the older living brother. Freud's death wish to Jung is that, like Julius, Jung is nothing but a doppelganger, or at best experienced as a phylogenetic memory.

Freud used Jung to explore the cultural unconscious. He needed a straw man to be the irrational one, while he took refuge in logical oedipal competition. Sacrificing Jung rather than himself, he denied his own instinctualized aggression. In the aftermath of their split, in what Henri Ellenberger (1970) termed a "creative illness," Jung dreamed he killed the hero Siegfried. Jung's *Redbook*, a half-life-long journal of creative psychotic regression reenacts Freud's damaging effect on his psyche (2012).

In the following months, Emma Jung, Carl's wife, begged Freud not to dismiss her husband; Freud placated her in private letters without acknowledging his role. On the other side of his primal ambivalence, Jung, Fliess, and Julius, all his symbiotes, were loved self-extensions, accompanying Freud through life. Like archaic leaders, he tried to keep his killed alive. Like Gilgamesh, Freud searched for a permanent cultural self, for a navel of associations still connected to his afterbirth, for a way to join Jung in a shared collective unconscious. Instead of mourning, heroes try to recreate a new social world.

FREUD'S CULTURAL EXPERTS: WILHELM WUNDT AND WILLIAM ROBERTSON SMITH

Gearing up for *Totem and Taboo*, Freud studied Wundt's *Elements of Folk Psychology*. Wundt, the founder of modern psychology, portrayed a hierarchy in worldwide culture:

> The striving to become a world religion ... is also a symptomatic mental phenomenon, paralleled externally by the extension of national States beyond the original limits set by the tribal unit. Corresponding to this extension, we find those reciprocal influences of cultural peoples in economic life, as well as in custom, art, and science, which give to human society its composite character, representing a combination of national with universally human elements. (Wundt, 1916, p. 10)

Freud looked up to Wundt with competitive ambivalence because Wundt was too close for comfort to Freud's desire to be the founder of psychological analysis. Nevertheless, in *Totem and Taboo*, he credited Wundt for understanding the origins of totemic culture. Wundt proposed that mental processes derive from internalized identifications through social interactions regulated by cultural evolution and its internalization through language. He outlined the stages of cultural development that Freud used:

1. Primal horde as primal herd
2. Totemic culture as mythic narratives of diverse animal groupings
3. Heroic individuals as founders of religious and military hierarchies
4. Nation States heading a hierarchy of clans, tribes, religions, and individual roles

Wundt described the evolution of these groupings from imitating animals in their survival mode, to gradually forming higher categories of identity that include animal gods, then human gods. Identity was conferred by culture over eras as individual leaders learned to use their power to form hierarchies of roles necessary for the survival of the group. The cultural hierarchy binds aggression, giving rise to feudal-type identity. One's place in social hierarchies is an identity marker for individuals and their kin.

William Robertson Smith's *Lectures on the Religion of the Semites* originally published in 1894, provided Freud with many of the fundamentals of his theory of the development of totems, taboos, and religion and so foreshadows some of Freud's cultural hypotheses. The extent of his reliance on Smith's position shows us how important it was for Freud to build on identifications with his own heroes. Just nine years older than Freud, Smith inspired Freud by how he used logic and honesty to respond to accusations of heresy by the conservative Scottish Free Church. Smith remained firm in his position and emerged as a well-known professor. Freud thought him a genius. While writing *The Interpretation of Dreams*, Freud dreamed of having his own theories accepted as a great discovery that would lead to the title of Professor Extraordinary.

Writing *Moses and Monotheism*, he followed Smith's reconstruction of Hebrew culture. Smith analyzed the cultural development of Semitic tribes described in the Old Testament, where Jehovah and other gods were believed to establish holy lands as places for survival. Rituals and rites were felt to imitate gods' presences as preservers of life—demonstrating what the gods expected followers to do. To be a member of the religious group, one had only to imitate the intention of the god. In this way, myths linked to the followers' phantasies. As tribes joined other tribes, each kept its earlier gods.

Freud added that rituals, as imitative actions, are linked to obsessive-compulsive symptoms. They are a fixation to the anal stage of development, for example, word and gesture, magic, and obedience training. Magic chants manifest obedience to the gods' power. Gradually the leader of the tribe, who communicated the will of the gods, was joined by shamans whose acts on behalf of the gods mediated people's anxiety about their leader's aggression. Shamans taught sacrifice as a way to appease the leader and the gods. Freud understood circumcision as such an act of appeasement. Appeasement and sacrifice were precursors of what Freud called the critical faculty or superego, because they were based on taboos, that is, rules of culture that cemented kinship ties and mitigated the force of the leader's—or demigod deities' aggression.

Smith described 800 BCE Biblical History when the Jews lived with other Semitic tribes under Assyrian despotism. Describing the formation of multiple tribal entities that lived under one king while following their own gods, Smith presents tribal unity as "an enemy to their enemies, a stranger to their strangers." He described how tribes affiliated with totem animals as gods: All creatures depend on what the gods provided for survival; hunter cultures often had totemic wolves; pastoral cultures had bulls. Forest animals were gods because of their connection with holy land that held survival resources.

When Jews or other Semitic tribes had to share survival with other groups, they developed the policy of the "inviolate guest." For instance, Jews are bound to open their houses to strangers to share a Passover meal. In order to survive it was necessary to recognize the rights of other groups of strangers to coexist— perhaps like animals at the watering hole. This cultural theme was important to multi-tribal, multiethnic cultures when raw aggression to the stranger had to be controlled. In the third and the fourth centuries when Rome was losing its empire to migrating groups from their north, churches were safe havens to mediate the survival of souls. As Smith put it, Mecca is also a place where the pilgrim can be a guest of god.

Like Smith, Freud believed gods arise as the extension of need, not as an extension of nature. Like prayer, rites convey the wish to impact physical reality. When gods are believed to animate holy places—the sources of survival—people gather there and live in a cultural center. Gods, as human and animal extensions, were imagined to be part of the physical world. Freud read in Smith that spirits were suggested to people by their dreams. Citing the core function of imagination in human relationships, Smith points out that transformational myths, poetry, and dreams are as essential to primary cultural process as gods are. Like gods, totem animals and imaginary creatures (griffins, sphinxes, and dragons) belong to a whole class of social creatures. *Jinns* are individual beasts that can mingle with the holy classes destroying survival possibilities.

Thus, taboos developed: Jews should avoid the evil, impure, and unclean. The idea of impurity, in addition to foreshadowing germ theory, also implies that objects are not what they seem; they have been degraded. Smith writes that positive taboos about what benefits all move toward moral order, while negative taboos move toward superstition.

NARCISSISTIC INDIVIDUATION: TAKING ON GOD'S ROLE

If a god's holy land is forest, pasture, steppes, springs, oases, agricultural valleys, high mountain tops where rain storms gather, or ocean waters where the vapors rise to make the clouds, then a person's hut or tabernacle is his holy land. He earns the social identity of *mine* by working the land, making it productive as gods do. That reminds me of the man who bought a decrepit house in a yard of weeds. After years of working on his property and adding a nice white picket fence, he stood by the gate as the minister walked by. The minster greeted him, by saying what a grand accomplishment he and god had performed in making that beautiful home and garden. The man answered that the minister should have seen it when it belonged to god alone.

Temples of worship acquire the holy quality where god appears. Public places have their origin in the sanctity of developing group morality. To this day the homeless can congregate in public places rather than in private ones, and so can people who want to make their voices heard about the morality of government. The agora, the market place where all are welcome, binds sectarian aggression. When terrorists choose to kill in a holy or public place they unbind aggression from cultural morality.

Freud often picked up on primal words that mean the opposite. In *The Uncanny* (1919), he investigated *heimlich* and *unheimlich*, remarking that the unfamiliar or strange can give rise to a frightening sense of unreality. Primal words reflect innate ambivalence, which plays out socially in kinship versus stranger. Smith had earlier analyzed *taboo*, *holy,* and *sacred* in their Semitic origins and found that these words all refer to moral *and* immoral aspects of their referents. In *Totem and Taboo*, Freud noted the double use of taboo as both sacred and unholy.

The idea of contamination shows primitive thinking in the sense that things are not what they seem; they can have the opposite quality. Holiness is catching by touching holy objects; unholy objects can also spread their quality by touch. In the fourth century, when Christians believed lepers had lost their souls, Constantine was miraculously cured of leprosy by converting to Christianity. But miracles can change valence from sacred to profane—curses can invoke negative miracles by unbinding aggression from morality. In that regard, religious forgiveness relieves breaches of rules and taboos. Freud thought of the dynamic unconscious as a flow between antithetical primary process and secondary process meanings. Conceivably, culture binds aggression with language by using primal words in the formation of moral concepts.

Chapter 6

Darwin through Freud's Eyes

Let's see how Darwin influenced Freud before studying *Totem and Taboo.* As we saw in Freud's relationship with Jung, Freud pictured himself as the leader of something like a primal horde of analysts; he felt that Alfred Adler and Carl Jung were among the band of brothers out to "kill" Chief Freud. Chapter 6 explores Freud's and Darwin's subjective childhood imagery as a source of their cultural theories:

> In 1912, I took up a conjecture of Darwin's to the effect that the primitive form of human society was that of a horde ruled over despotically by a powerful male. I attempted to show that the fortunes of this horde left indestructible traces upon the history of human descent; and, especially, that the development of totemism, which comprises in itself the beginnings of religion, morality, and social organization, is connected with the killing of the chief by violence and the transformation of the paternal horde into a community of brothers. (Freud, Totem and Taboo: The Group and the Primal Horde, 2013, p. 122)

Darwin's transformation of his own life experience to his theories of evolution shows how co-extensive fantasy and scientific logic are. Freud's reconstruction of cultural stages in *Totem and Taboo* showed him that his own childhood fantasies were recapitulated in the childhood of culture. In the following section I will recapitulate Freud's core early childhood memories to show how they entered all of his developmental fantasies at the core of his life experience, even making their way into his rendition of psychoanalysis.

FREUD'S ARCHAIC UNCONSCIOUS

Freud believed that myths like those of Oedipus and Moses, putting the infant hero in a basket as a foundling, was a code way of saying that infanticide was practiced by fathers to maintain their power and privilege and to forestall patricide. Researching the dynamic culture of primal hordes in 1913, at age 57, Freud felt oedipal competition in leading his psychoanalytic movement. In *On the History of the Psychoanalytic Movement* published in 1914, conflicted about removing Jung from his psychoanalytic movement, he writes:

> I was no longer young … I felt oppressed by the thought that the duty of the leader should fall on me so late in life. … I felt the need of transferring this authority to a younger man, who would then as a matter of course take my place after my death. This man could only be C.G. Jung … he seemed ready to enter into friendly relations with me and for my sake to give up certain racial prejudices which he had previously permitted himself. … I had lighted upon a person who was incapable of tolerating the authority of another … whose energies were relentlessly devoted to … his own interests. (Freud, p. 43)

Freud alluded to anti-Semitism. He wanted psychoanalysis to be international and interdenominational; like Darwin he thought diverse cultures survived better. But he saw Jung as seeking power, women, money, and influence, as entitlements of becoming the psychoanalytic leader. In Freud's dynamic cultural unconscious he disposed of Jung as he had disposed of Julius, the younger son—but also as he had disposed of the contender to his throne. Freud's confused primal ambivalence shifted like an electron in two orbital positions at once, identified with his Jewish heritage and his Christian one-eyed doctor.

In *Totem and Taboo,* Freud cites James J. Atkinson's *Primal Law* in which the primal horde—*The Cyclopean family*—develops incest laws to prevent infanticide by despotic fathers who eat their sons, and parricide by brothers to take his place—until mothers bound that raw aggression by making their youngest son a leader. Consequently tribal rules developed as taboos against incest and cannibalism. Freud did not have a despotic father, but his pervasive imagery like the Cyclops in Odysseus's travels, is a one-eyed monster who cannibalizes all men he sees.

We will linger for two paragraphs in *The Interpretation of Dreams,* Freud's memoire of poetic associations made scientific. In his *My Son, the Myops* dream, Freud associates to imagery from his childhood revealing dynamics and conflicts entering his discovery of psychoanalysis. A caretaker nun holds back tears when she hands two brothers to their father (pp. 441–444). Freud recognized his older son, the younger is faceless. He associates to a professor whose son had an inflammation of one eye that cleared up, but the other eye might get worse. (Freud also had a one-eyed history professor.) Freud saw Theodore Herzl's play about Jewish doom the evening before the dream: migrations without a homeland. Reliving his early childhood in the dream he envisions Monika mourning Julius. After Sigismund's dysentery cleared up, Julius's ended in death. Then Sigismund's extended family entered its diaspora.

From Myops to Cyclops referred to walking with Fliess at Easter/Passover the year before the dream in a vacation in Breslau (where three-year-old Sigismund had gone with his mother on a train, beginning their migration to Vienna). In association to the same dream Freud remembers Fliess talking about bilateral symmetry, "*If we had an eye in the middle of our foreheads like a Cyclops ...*" (Freud, p. 443)—just then a girl asked them for directions they did not know. Then seeing a sign—*Dr. Herode's consulting hours*—Fliess joked that the girl should be careful about whom she asks for directions: Christians might think Jews used Christian children's blood to make matzos. Freud recapitulates his organic brain theories about how conflicts form by instinctual oppositions at the time of his dreaming. Thus, the two sides of the brain represent bilateral opposites: living or dying, staying or going, and knowing or not knowing what is happening. "My concern about one-sidedness had more than one meaning. ... After the child had turned to *one side* to say farewell words, he turned to the *other side* to say the contrary ... *as though he was acting with due attention to bilateral symmetry!*" (Freud, p. 444).

Freud's core ambivalent imagery repeats his splitting: Julius died, but he was still part of Freud's own mind as a double. Moreover, in regard to Freud's splitting between a Jewish father and a Christian father figure, on the eve of his father's funeral he had dreamed:

> You are requested to close the eyes, or
> You are requested to close an eye. (The Interpretation of Dreams, p. 317)

But let's follow his imagery into another dream segment on the night before his father's burial in October 1897 when he was immersed in recording his own dreams about how the mind and brain create emotional life. The key to Freud's preoedipal confusion, his identity splitting, and the imagery of his dreams is contained in letter #71 to his friend Fliess (Freud, 1897). In the previous letter he confessed that he had greeted Julius, "barely a year younger" and dying after a few months, with "ill-wishes." He added a dream associated to his "old" nurse who taught him "how to live": he saw a skull of a young pig (a dead baby at the breast). The letter brims with excitement about discovering *Oedipus Rex*—or was it pre-*Oedipus Rex.*

As we will see, Sigismund's unknowable source of guilt over his brother Julius's death became buried under his theory of the Oedipus Complex. Letter #71 condenses his confused young childhood ideas of the mysteries of birth and death with facts of life and death. Freud says that it is universal for children to fall in love with their mother and to be jealous of their father—earlier in life in cases of hysteria like his. In Greek theatre each audience member recoils in horror from realizing his or her own infantile state. He tells Fliess that Hamlet's obsessive confusion about revenging his father's murder came from a distorted oedipal theme. But Freud's self-analysis of his own young childhood was swayed by preoedipal thinking. Birth, death, sex, and murder condense into primordial fantasy, everyone's repository, everyone's navel.

DARWIN'S PERSONAL EVOLUTION

When our shamans, storytellers, narrators, memoirists give us the legacy of their life, they transform their life experience into reflections that may enter the holistic culture ongoing after any death. Trauma and loss come to everyone's life exogenously, but endogenously, from innate cultural heredity, they come during the inexorable passages from one stage of life to the next. Each of us has to deal with his or her unconscious cultural dynamics roiling the calm of every-day life. It is up to our story-tellers, artists, and creative thinkers to interpret the effects of cultural geist (spirits) on present survival.

Just as story-tellers use creative fantasies to find their material, scientists like Freud and Darwin analyze their unconscious to find the direction of their research. Here I'll offer a perspective on Darwin's unconscious conflicts as they permeated his theories and provided Freud with a template for understanding culture. Freud used Darwin's ideas to investigate the formation of culture, especially those ideas derived from Darwin's identification with his mother. Freud believed, as Darwin did, that innate bisexuality energized male contributions to cultural progress. If Darwin's life experience informed his theories of evolutionary survival and *The Descent of Man*, let's study how he did it.

My take on Darwin's life is based on his *Autobiography* (1887). His concern with death as a sign of hereditary weakness began with his mother's death when he was eight years. His father, a doctor, decried the weakness of his women patients who did not realize how their irrational approach to life doomed their health. He expected Charles, like his older son, to follow in his footsteps and become a doctor. But Charles seemed irrational and sentimental, with a compulsive nature and anxiety that broke through more the older he got. Dangerous images came into thought compulsively, against his will. Thus, in *The Descent of Man* (1871) he concluded that the highest intellectual moral development in an individual is the internalization of cultural control over one's thoughts. Perhaps, as well as representing the ineffable truth of culture's effect on identity, the control represented his ambivalence toward his father.

After his mother died of a stomach ailment, little love was available to him; his father forbade mentioning her. His two older sisters reinforced the father's edict. Not allowed to talk about his mother or to express emotional pain, he was sent away to school, where he made many friends and came to look up to men other than his father, especially Josiah Wedgewood, his mother's brother. He transferred his male social bond to this uncle. He became a collector of birds' eggs and other forms of life; and developed a passion for killing birds. His talent was observing specimens and noticing any deviation from their species. As a boy collecting beetles he could recognize a new variant on sight: what I call a RB capacity to conceptualize variations in a whole gestalt. Identifying with his father's control, but feeling deep sympathy and empathy with his mother's abdominal pain, he transformed a drive for exploring novelty into a passion for specimens, facts, and theory, all to hold onto his mother's image as central to his identity.

Like the ancients, we respond to death by identifying with the person who died; and facing the prospect of our own death we leave our culturally conceived soul to the young as legacy. Darwin opens his *Autobiography* with an invitation to his kin and a message to all readers, stating that the account of himself is the legacy of a dead man meant to ease the pain of his family by knowing his love for them all. His sparse childhood recollections are nuclei of suppressed emotions of longing. He remembers bathing in the sea with his mother when he was four, seeing her dead and laid out in black, stealing sweets, and on occasion stealing puppies' love from their masters. Carefully defying his father, he felt that taking any emotional sustenance was stealing. Perhaps, in his oedipal fantasy, he stole his mother back from his father, a physician who could not save his own wife. He invites his readers to take his spirit with them through their life voyage, just as he took his mother's spirit on his.

Darwin stole the evidence of mothering by taking just one egg from each nest. Collecting living things and residual signs of their life excited him. By extending latency-age collecting into adult life, he manifested an obsessive-compulsive character structure, replete with neurotic anxiety and ambivalence. At 15, hunting with Uncle Josiah, Charles became an impassioned bird shooter. Shivering with ecstasy, as he said, shooting birds out of the sky, with retrieving dogs as a self-extension, he described every trophy bird with the words that might have the magic power to bring his mother back.

Instead of studying medicine at Edinburgh University when he was 16, he studied natural history. He brought the dead back to virtual life by learning taxidermy from John Edmonstone, a freed black slave. Edmonstone informed Darwin that in South American rain forests he could find new species actually coming into being, so in his five-year voyage as the *Beagle's* naturalist,

Darwin searched there for those *creative centers.* Conceiving the land mass of South America to be still rising from the sea on platforms like the Galapagos Islands, he searched for new species arising from the ocean.

Imagine his excitement, shooting finches out of the Galapagos sky, hoping they were new species. Identifying new species arising from the old—still containing the old—seems to me an unconscious fantasy of keeping his mother alive in himself. Collecting so many specimens of so many species under so many conditions in his five-year voyage, Darwin gave himself decades of hard work ahead, sorting, theorizing, and writing.

When he returned from the *Beagle* at age 30, with admiration from naturalists and further work cut out for him, he had a decision to make about marriage. Courting his first cousin Emma, Josiah's daughter, Charles made out a list why and why not to marry: dogs are better companions; a bachelor has more time to collect specimens and to write—but who will take care of him? Marrying a woman representing his mother caused him to hesitate. Would she too die?

After a year, he made up his mind and married Emma, who gave him motherly care, companionship, and 10 children. But Charles had abdominal anxiety symptoms that felt like symptoms his mother suffered before death. They worsened during Emma's first pregnancy, but lessened after the second, when Annie was born. Like a little mother to Charles, Annie was the one who helped him recover from his anxiety attacks. When she fell ill at age 10, perhaps of TB, Charles took her to his own physician, Dr. Gully, who had sent him to a clairvoyant in the past when hydrotherapy failed to relieve him of his symptoms. Perhaps a clairvoyant could keep Annie alive—but to no avail.

Darwin's irrationality shows how life experience generates wishes for restitution: theories however misbegotten in fantasy that seek compensation. Fantasies influence new experiences of events that find their way into reflection on identity, which is the innate brain capacity that feels like indwelling otherness—and culture forever. Cognitive reflection judges the relevance of events, relationships, goals, and states of identity, experience meant for communication to others as well, sometimes as a legacy to those you love. Mature narcissists feel they are contributing their reflection to the general good of those they care about. In their creativity, mature narcissists attempt to become heroes who restore what has been lost, which was felt in Darwin's case to be their gift to humanity and to being part of that humanity.

SCIENTIFIC FANTASIES CAN COME TRUE FOR A WHILE

Like Freud, Darwin, had fantasies of bringing back revenants as compensation for death and loss. Darwin began with a belief in god's creation of separate species by design; but after studying the embryological similarity of all vertebrates in human fetal development, which showed human descent as a vertebrate, he called himself an agnostic. Like Freud, he never completely relinquished the design idea. Freud accepted Darwin's idea that using our brains creatively facilitated survival and inheritance. Brain evolution and bipedal posture freeing the hands, gave rise to inherent thinking patterns that contributed to everyone's survival. Freud and Fliess believed that bipedal posture brought the nasal turbinates, as sexual organs of discharge, away from the genitals and anus, lessening instinct-triggering smell (pheromones) that often pervade mammalian thinking.

Freud followed Darwin's surmise that archaic humans were closer to our animal origins. Much of what archaic people projected onto totem animal gods transferred into culture, first as religious beliefs, then as scientific beliefs. *The Descent of Man* expresses Darwin's misgivings about marriages like his that are too close in the hereditary chain, causing weakness of the offspring and early death. He was fascinated by how the male and female of a species vary the same embryological origins of the reproductive organs. Identifying with his mother while writing *Descent*, he repeatedly describes the residual capacity of male breasts to give milk. *Descent* cites mammalian faculties of reason and sympathy as foundations of human cultural evolution. He understood the moral sense as a social instinct that mammals in particular developed through natural selection. His mother's love pervaded his wishes. Darwin showed his sympathetic altruism in his famous last words of his *Autobiography*, spoken as he died in his wife's arms: "I am not at all afraid of death. Remember what a good wife you have been to me. Tell all my children to remember how good they have been to me" (Darwin, 1887).

FREUD'S IDENTIFICATION WITH CHARLES DARWIN

With intense interest and excitement, Freud absorbed many concepts from *The Descent of Man*. Darwin distinguished separate instincts for individual and species survival. He was known for his views on competition favoring survival of the fittest when Freud was growing up. Many species show male and female fitness when assertive male display wins female mating—like human individuated narcissism. Darwin's theory of the primal bisexuality of the species had pervaded modernist cultural thinking. Freud's version of innate bisexuality posits sexual mating and oedipal motivation as well as homosexual sublimation for the sake of cultural progression; while the modernist intellectual community believed Darwin's theory of atavistic bisexuality as a form of throwback toward a primal animal state. Jews were thought to characteristically manifest atavism. Freud followed Darwin's lead in *The*

Descent of Man that vertebrates developed by differentiating earlier evolved hermaphroditic sexual glands.

Darwin wrote of the prostate gland as sharing the embryological origins of the uterus.

Inspired by Darwin's ideas when he was 21, Freud brashly submitted a research proposal and was given a grant to investigate histological sources of the bisexuality of eels. He went to cosmopolitan Trieste where there were eels—but found no evidence of ovaries in male eels. In fact, eels change their sexual organs from year to year. Freud's follower Ferenczi noted that the embryonic gills in the human fetus indicated human descent from the sea, which led him to infer that primal feelings include the uterine experience of being awash in origins. Freud inferred that chemical hormonal substances distinguished male from female sexuality, balanced differently.

WHERE DOES IDENTIFICATION COME FROM?

People share biological instincts that energize a consciousness shaped by culture. The primal horde family group with its familiar cohesion is symbiotic. Freud posed body identity in terms of self as the source of action, and others as the target of satisfaction or danger. Following Freud, we can conclude that identity progress from infancy to young childhood recapitulates moving from the primal horde to the tribe. The passage out of infant all-or-nothing, love-or-hate, experience into young childhood establishes identity ambivalence: either you join the tribal totem or you become taboo. Objectified by others as a separate member, the young child clinging to symbiotic wishes feels the loss of having been a baby. That feeling of past identity merges with tribal collective memory: the dead live on as mythical self-extensions: phantasies.

Older members of the tribe know that the dead have been absorbed into the structure of culture that the young child enters. The structure of cultural reality feels like a secret that has to be decoded, bit by verbal, cognitive, and affective bit, as magical revelations about how to imitate to find a role. Maintaining continuity of the tribe's past entail a belief in totems that control feelings of tribal identity. Learning by imitation and identification to be a member of the tribe feels like entering new reality. Freud described the unbearable feeling transition to a new social reality of being alone as a *universal primal scene*.

CULTURE HAS ITS PLACE IN THE BRAIN

Here, let's add brain network perspectives to psychoanalytic metapsychology. Medial networks specialize in social processing, which supports Freud's view of a top-down linking of mental and cultural organization with bottom-up instinct. From that perspective, neuromodulators connect drive sources in the midbrain to organizers of thinking and feeling. Motivation for pleasure or to relieve emotional pain differs conceptually from instinct. From a top-down mental perspective we find LB libidinal sources for maintaining the internal milieu and sexual reproduction, and RB aggressive sources for protection from physical or social damage. Culture is imbued with a sense of aggression greater than any person's.

Corresponding to a neuroscience perspective, Freud wanted to understand the mind from a cultural, top-down and a biological, bottom-up perspective. He proposed that identification is the core mechanism of instinctual binding by culture, a means of building mental tendencies into brain process. Feeling others as the source of pleasure, for example forming a *cathexis*, transforms brain process to mental process. Identification shows how to imitate others to get what they have or their protection. So cathexis and imitation are different forms of relating. Becoming like the other brings otherness into the self as a willful, LB, capacity to do what the other can do. The sense of self as the source of action always begins as something borrowed from others—the wish behind the construction of fantasy. When others show you how to behave to get what you want while staying safe, their judgment enters ego identity as superego.

I propose that executive identity continually remodels character structure by adding new identifications in the medial PFC to link modules of VMPFC-self and DMPFC-otherness. Daily social experience triggers daily neural adaptation. We can see how this works by considering changes in neural adaptation during PTSD. When abused children grow up they often show brain changes consistent with PTSD, including amygdalar and hippocampal shrinkage that limits new memory formation, and cortisol instability in response to imagery associated with past or new trauma. Compared to normal children, gray matter in the abused child's VMPFC over-expands in the years following abuse, while DMPFC gray matter contracts, which corresponds to identifying with the seducer/aggressor (Richert, Carrion, and Karchemskiy, 2006). Given the context I've outlined for the mental function of these networks, I think that VMPFC-self increases at the expense of DMPFC-otherness. Further research (Carrion et al., 2009) showed increased left VMPFC gray matter in abused children compared to others.

Now, let's infer how identity changes adapt to network changes in PTSD in children. Being the target of adult abuse causes psychological identification with the abuser. Ferenczi's classic *Confusion of Tongues between Adults and the Child*, 1933, presents a psychoanalytic perspective on sexual abuse: The child identifies with the abuser by feeling responsible for the sexual activity. He or she incorporates adult motives as adult stimulation encompasses the child's self, turning others into potential sexual partners. As the abuser becomes familiar, the stranger quality dissipates in a pathological identification. Assimilating the instinctual tendencies of the abuser, over-stimulated children may become adult sexual abusers. Ferenczi said the guilt as well as the sexual and/or aggression of the perpetrator is also internalized by the victim.

INHERITING CULTURE IN CHILDHOOD

In *Group Psychology and the Analysis of the Ego,* Freud writes:

> A path leads from identification by way of imitation to empathy. … The study of such identifications … led Robertson Smith … to the surprising discovery that they rest upon the acknowledgement of the possession of a common substance … created by a meal eaten in common. … (Freud, 1921, p. 110, Footnote 2)

He concluded that group identification with the leader of a tribal clan rested on a collective phantasy of ancestors who incorporated the leader of a more primal horde into their cultural taboos and customs. The incorporation is ritualized by a shaman in a sacrificial feast of a totem corpse. In Freud's reconstruction, a primal horde of brothers, jealous of the father's privileges, killed him, cut his body into pieces, and shared the feast with one another. Remorse came when antagonism among the brothers threatened social survival. To restore order over prolonged collective time, the mother's favorite, her youngest son, became the new leader. The new leader added taboos that bound libidinal and aggressive instincts, by sublimating them in song, dance, customs, and ceremonies.

Freud proposed that the family's cultural training organizes the child's preoedipal development. In oedipal development, states of imagination organized toward wishes for future gratification, reorganizes psychic identity. Imagination has a future quality; but so does fear. As the sense of identity time becomes continuous during latency, the anxiety generated by fear for body integrity represses the fantasies, which give rise to a latency organized sense of culture enveloping the whole family. Freud imagined that oedipal crimes over millennia triggered the evolution of oedipal fantasy detached from action through latency control. His theory conforms to brain-based development theories: (1) Luria's neocortical *second signal system* overlays earlier development, (2) Piaget's imaginative parallel-play stage yields to cognitive development, and (3) Vygotsky's internalized social language reorganizes thinking. Fantasy seeks future identity pleasure, as ambitious daydreams are worked through by maturation of the DMN throughout latency (Faire et al., 2008).

Compared to other mammalian development, humans evolved more narcissistic individuation. When a herd leader dies, another male develops hormonal changes, including higher testosterone, cortisone, and norepinephrine, to prepare him for power and leadership. For human brothers to fantasize they could change their group role and kill the leader was an evolutionary breakthrough. It is one thing to play at butting heads to develop alpha spirit among the male animals growing up together, but another thing entirely to imagine changing identity—from being part of the group to becoming its leader—by killing the father of the horde. Oedipal-based fantasy expands to latency fantasies of cultural ambition, and then to adolescent fantasies of sexual exploration. Identity-changing fantasies promote changes in social identity as they prelude to later life passages.

FREUD'S FAMILY ROMANCE FANTASY

From ages 6–10, children build a reflective sense of their indwelling social world, which includes an idealized, utopian, progressive culture—or perhaps a gang. Sometimes, the family romance heroes or heroines aggrandize their place in their family's social history. Freud wrote that those who leave their mark on culture develop a determination to do so at an early age. In his latency, Freud worshipped Shakespeare, Cervantes, and Goethe, European narrators of cultural development. He felt they were progenitors working to create a progressive world culture.

Freud believed he was his mother's hero. In *The Interpretation of Dreams,* in association to a dream, he says being born wrapped in a caul at birth induced an old peasant woman (Monika?) to foretell his greatness. Associating further, he recalls being in the Prater Park in Vienna with his parents when he was 11 or 12; and a wandering poet foretold he would become a cabinet minister (Freud, 1900, pp. 191–192). Freud often expressed his narcissistic wish to become a *professor extraordinaire* despite being a Jew. He sought influential friends who, like his alternative parental imagoes, were Christians. But he turned his narcissism toward self-objects when he adopted younger men to become literary helpers and muses, only to reject them when they followed their own ambitions.

Just as Freud treated Jung as an afterbirth to his own greatness, so Otto Rank was an extended self-figure whom Freud used as his muse in exploring the myth of the hero's birth. When Rank was 21 and working as a locksmith, after reading *The Interpretation of Dreams*, he wrote a monograph about the artist from the perspective of a transformative sexual fantasy, which he used as his entrée into Freud's world. Freud's narcissistic self-love attached to Rank fulfilled the sublimated homosexual fantasies of each of them. Freud's

Family Romances paper (1909) was first published by Otto Rank within *The Myth of the Birth of the Hero* (1904). Although it was Freud's concept of *family romance fantasy* as the origin of the oedipal hero, he allowed his poor young Jewish protégé to publish *Family Romances* as if it were his own. Freud knew that sublimated bonds among men often included self-transference, loving a younger version of the self.

In the *Family Romances* paper, Freud (or Rank) discusses the latency child's tendency to deny parentage through a fantasy of having been adopted at birth by a poor but nurturing family, while imagining his real father was a hero. A family romance fantasy loosens attachment to parents and family so the child can find narcissistic autonomy in the social world. Rank's family was too poor to send him to college, so Freud helped him enter and graduate from a PhD program at the University of Vienna. Then, Rank changed his surname from the Jewish *Rosenfeld* to Rank—a more powerful German variant of male strength than *a field of roses*. Rank was the only paid member of Freud's inner circle.

The family romance fantasy is often transformed in puberty into an identity-organizing oedipal fantasy that pervades ambitious daydreams of conquest; it is most marked in talented—or paranoid—people who wish to see themselves as exceptional heroes. In adolescence, the alternative-parents theme sometimes shifts to a belief that one's mother had an affair with a prestigious person—the real father. In his *Family Romances* paper Freud writes that the mother often enters the boy's oedipal, virgin/whore phantasy:

> ... a boy is far more inclined to feel hostile impulses toward his father than toward his mother and has a far more intense desire to get free from *him* than from *her*. In this respect the imagination of girls is apt to show itself much weaker. These consciously remembered mental impulses of childhood embody the factor which enables us to understand the nature of myths. (Freud, 1909, p. 238)

A MODERN FAIRY TALE

But the family romance phantasy and its modern variants also enrich female narcissistic creativity. I treated a woman for several years for intense worry over her stepbrother's future. Her mother had been a 16-year-old child of an impoverished mother whose husband had died. A Greek sailor came to the Mediterranean island where they lived, swept the girl off her feet and her Island, married her, and took her to the U.S. Midwest to farm. But he had tertiary syphilis and died in the second spring of their stay, leaving my patient as a baby starving in the late winter with her young mother. Her mother had a distant relative in Idaho, who took them in. There, she married a man who treated my patient as his own child. As she grew up her stepfather would stimulate her clitoris on many nights, when her mother was pregnant, but he never raped her. Her mother had a boy she helped to raise, feeling responsible for him when her mother died a year before she began therapy.

During her therapy she married a divorced man who had five children she helped to raise, while working as a dental technician. Her compassion for her mother and brother transferred to caring for her husband's children. Two years later, she returned for a second bout of therapy. She wanted to make jewelry. Amazing to me, she had talent I had not known. It became clear in further analyzing her that when her stepfather stimulated her, he would leave her full of desire, wishing he would go on until—she did not know what. The feeling of seeking an orgasm became attached to the wish to make a beautiful jewel that would finally satisfy her. After the interpretation was made within the scope of father transference to me, she worked it through by organizing a workshop for making jewelry, with four women working under her direction. Her work was so lovely that a vice president's wife wore her earrings to the inauguration. Then she bought a dilapidated stable on Martha's Vineyard that she had redone for her family's vacations.

Her living Fairy-Tale thrived on knowledge of how her own narcissism could flourish when her sexual pleasure was sublimated to pursuing other's narcissistic gratification. She understood how to influence high-end fashion people by gratifying them with gifts of jewelry. In the realm of fancy people, creativity is valued as self-extending group fetishes. My patient's family romance fantasy of being saved, as her mother had been save from poverty by the sailor, was gratified by a stepfather who added excitement and survival security, becoming a sexual imago: a ladder for her development.

FREUD'S IDENTIFICATION WITH MOSES—AND OTHER MUSES

After Freud wrote *Totem and Taboo*, and disposed of Jung and Adler, he identified with Moses as the bringer of morality to the psychoanalytic tribe. For 17 consecutive days on a visit to Rome in 1913, girding his loins as leader of the psychoanalytic movement, he studied Michelangelo's *Statue of Moses: The Ten Commandments* are slipping out from under Moses's arm, while he turns from staring into the radiant eye of god to glaring at the Jews down below who are worshiping false idols in sexual abandon.

Freud's followers were worshipping other beliefs than his psychoanalysis. Freud did not know if he had the right to identify with Moses as the giver of laws. And so he was driven day after day to the great statue personifying conscience. Freud did not wish to publish the paper under his own name. His family romance fantasy was reaching out too far and too evidentially in his identification with Moses. In *The Moses of Michelangelo* Freud wrote:

> How often have I mounted the steep steps from the unlovely Corso Cavour to the lonely piazza where the deserted church stands, and have essayed to support the angry scorn of the hero's glance! Sometimes I have crept cautiously out of the half-gloom of the interior as though I myself belonged to the mob upon whom his eye is turned—the mob which can hold fast no conviction, which has neither faith nor patience, and which rejoices when it has regained its illusory idols. (Freud, S., 1914, p. 213)

Freud is writing in a literary style, wondering if he is entitled to be an individual creator. He imagines himself as one of a still primal, undeserving mob. He climbs the steps to the statue, like Moses climbing the mountain to meet the god of morality, the superego. He characterizes the Jews as a licentious, archaic group. Freud had already disposed of Julius, Fliess, Adler, Jung, and others blocking his search for morality and truth.

Who was he to lead his psychoanalytic movement? What gave him the moral authority to decide who lived and who died culturally, as a member of his group? His narcissism had been in full bloom as his mother's favorite, when it should have been Julius. How many of his associates had already died in vain pursuit of greatness? How much of that was his fault? First he seduced them, and then abandoned them: *non vivit.* Was he immoral by insisting that his sisters had to be quiet and not practice their music, while he studied, burning expensive midnight oil? Did he understand women's contributions to culture? Did his family romance lead him out of Jewish morality as a godless leader?

Years later, in Part 1 of *Moses and Monotheism,* Freud reconstructed the Moses myth, viewing him as being born an Egyptian, not a Jew. He demoted Rank, saying he got him to write about the birth of the hero, as a family romance myth accounting for many of the world's culture-founding heroes (Freud, 1939, p. 10). He says the Moses myth of the Jewish mother floating her infant son in a basket down the Nile to save him from the Pharaoh's edict to kill Jewish firstborn male infants imbedded in the Passover ceremony reverses that. Instead, in a syndrome of revenge and retribution the Jewish god passes over Jewish houses and kills the Egyptian first born. Freud's ambivalence about his right to lead vacillates in his writer's ego identity, among imagoes of the one-eyed doctor, the vulnerable father, the first born son, the dead younger son, and the mother who loves him most of all. Yet, if the Christian (or Egyptian) has the power, the Jew is merely atavistic.

Not until he was a very old and critically ill Jew on diaspora to London to escape the Nazis did his guilt move him toward a goddess culture. Finally, in writing and rewriting *Moses and Monotheism,* he wavered, recasting Moses's role as the Egyptian leader of a sect of Egyptian followers who believed in one unifying sun-god, a tribe called Jews, as they became a maternal goddess culture. He still pondered whether goddess cultures occurred after father-leaders were cannibalized. The remaining brothers could never trust one another until they followed the youngest son, Joseph, the mother's real favorite or was it his youngest daughter, Anna who ruled? In the following chapters, we will see what Freud went through in his position of the leader before giving the reins to Anna.

Chapter 7

From the Primal Horde to the Primal Scene

Chapter 7 studies *Totem and Taboo* (Freud, 1913), his basic work on cultural origins. Freud reviewed the implications of that work in *On the History of the Psycho-analytic Movement:*

> I have made an attempt to deal with … social anthropology in the light of analysis; this line of investigation leads directly to the origins of the most important institutions of our civilization, of the structure of the state, of morality and religion, and … of the prohibition against incest and of conscience. It is no doubt too early to decide how far the conclusions thus reached will be able to withstand criticism. (Freud, 1917, p. 37)

Let me say at the outset that in its reconstruction of cultural development, I think it fails to treat aggression as a core instinct equal to libido. Let me also say that the term aggression can be used in opposing hemispheric ways: aggression as a determination to find pleasure, and aggression as a response to either you or your kin being threatened. Either side of aggression can mobilize primal rage. Each childhood stage develops de-instinctualized, bound aggression and bound libido. Freud accounted for aggression in conceiving love/hate as innate ambivalence, but the quality of hate differs from the quality of dislike or anger in childhood ambivalence, and from the wish to eliminate oedipal competitors. Sadism and fairness show unbound versus bound aggression developing in latency.

Freud conceived *universal,* traumatic passages accounting for stages of childhood corresponding to stages of cultural evolution: being born into infancy, raging at young childhood reality, detaching from wishful oedipal fantasies to shut down anxiety. Each passage binds sexual motives and aggressive affects that enable practical thinking. Each passage causes social changes in identity experience.

Freud noted that a totem animal was a source of cultural identity for all tribes. Sometimes a totem could be incorporated whole, like what the caribou provides the Inuit people: meat and fat for metabolic sustenance; oil for light; hides for clothes, dwellings, and coverings; sinews for sewing; and horns for tools. Other totem animals, meat-eaters like wolves, showed a tribe how to hunt. Totemic horses were used by Asian Steppe nomadic tribes' attacks; their warriors drank fermented mare's milk for energy and stamina. For the Shoshone tribes of North America the wolf totem was a demigod, creator, sometimes becoming a legendary man—the leader's inspiration. Every member of a tribe would identify with its totem as a sign of membership.

TOTEM AND TABOO: PART I: CULT PRECEDES CULTURE

> Linguistic usage in these Australian tribes exhibits a peculiarity … relevant here. For the terms used by them to express the degree of kinship do not denote a relation between two individuals but between an individual and a group. (Freud, 1913, p. 6)

Culture encompasses a group identity fueled by each person's biological energy. Mothers tend to provide satisfaction; fathers, safety and leadership. A regressive cult as opposed to a primal cult arises when the leader's self-privilege—primary narcissism—overrides his commitment to cultural survival. A primal horde was vulnerable to its father's/leader's primal narcissism, if that narcissism was too great; the inflexible horde would likely fail to survive. Studying that tendency, Freud noted that later in history, kings would glorify themselves as leaders, using custom and taboo to claim ownership of land that once was the habitat of their totemic gods. But, nomadic leaders were not so much interested in owning land as in controlling access to grazing passage for attacks on settled cultures.

By requiring exogamy to prevent incest, primal hordes divided into subsections of a larger tribe, but they maintained obedience to one leader who embodied group culture. Freud treats tribal culture as a group phenomenon, in which the leader demonstrates how the totem animal will insure the tribe's survival. Kinship within each tribal subsection is structured by incest taboos—rules that bind libidinal and aggressive relationships. As Freud saw it, proscribing incest was the way to structure tribal culture, preventing conflict among male kin contending for access to libidinal satisfaction. Tribes developed exogamy rules to bring an outside view to each new marriage horde that helped bind the father's /leader's narcissism. Thus, marriage partners often had a different totem, which enhances Darwin's view that incest leads to biological weakness.

As multiple tribes formed federations cultural rules and roles had to differentiate. The equality of leaders enhanced survival when the totem-way loses reliability. Councils conciliate. In highlighting sublimated male bonding as the means for diplomatically building culture, I think Freud underestimated the value of male aggression in exploring novelty—as in the case of Gilgamesh. From a neuroscience vantage of using the top-down neuronal workspace for figuring out policies that work, cultures evolved using anterior prefrontal cortex to process their own reflective hierarchy of cultural generalizations. LB hubs bind libido for future pleasure; RB hubs bind aggression with conciliatory responses.

Sexual and aggressive drives develop in different degrees in particular cultures. For example, the sedentary culture in Uruk, familiar to us from the *Epic of Gilgamesh*, showed the wisdom of sublimating libidinal narcissism. On the other hand, Genghis Khan's nomadic culture developed an aggressive military hierarchy parasitic on more sedentary cultures. Freud did not write very much about how more aggressive cultures formed their structures for survival and how those structures became incorporated in other cultures. Trade and religious influences gradually brought diverse cultures into a *modus vivendi.* In Uruk, the vast doors of the market place kept aggression outside, or by opening up to trade, controlled the aggressive designs of other cultures. Pastoral cultures, like post-diaspora Jewish tribes from Egypt, moved back and forth seasonally with their herds, between more nomadic and more sedentary peoples. On the Asian steppes, Buddhist teachers taught nomad leaders how to sooth high stress levels with relaxation techniques learned from Hindu culture. Eventually the more aggressive nomad cultures kept trade routes open along the Silk Road (Grousset, 1971).

ESSAY II: *TABOO AND EMOTIONAL AMBIVALENCE*

Culture finds its way into the habits of the body in order to subordinate individuals to its rule. Freud treats *taboo* as a primal word that manifests outside control inside the body. It is a form of aggressive binding through a shared belief in a totemic god of survival. For Freud, taboo evoked primal instincts: life and death, sex and aggression. Taboo is dangerously close to god. Culture binds the mind with taboo. Stealing omnipotence by violating taboo brings on death. Freud discovered the correspondence between taboos and the signs, symptoms, and superstitions of obsessional neurotic patients. Taboo prohibitions are the bane of an obsessive's existence.

Cultural laws emanate from the leader's *manna,* or power—the power of his words to control unconscious ambivalence, as in "Do not take God's name in vain!" The source of *manna* in the universal passage from infancy to young childhood is the power of adults to control reality with acts, words, and aggression. "Don't touch!" The power is in the words, gestures, examples, and punishment. When the young child or some child defies manna the tribe's aggression may end in abandonment. Even thinking about violating taboos causes anxiety—manna power inside the body is an aggressive demon. The angel of infancy gives way to the devil of young childhood.

In Freud's theory cultural structure determines the dynamics of ambivalence. If *condensation*—merging self and other in wishful identifications—is the infant mode of libidinal unconscious processing, then *displacement* is the young childhood movement of aggression from one object to another. From my neuroscience perspective, in infancy and young childhood the orbital frontal cortex (OFC) conditions social experience as either satisfying or emotionally painful, relaying the data to the VMPFC-self. Forever after, OFC bias precedes other forms of social responses. What is familiar feels satisfying, what is unfamiliar feels potentially aggressive.

Freud picked up on Wundt's concept of taboo as joined feelings of veneration and horror, the emotionally ambivalent essence of obedience to the leader. But for the young child and for young culture, before any idea of autonomy kicks in, the follower can only imitate the leader, using the magic of word and gesture to capture the unknown spirits that can inhabit their body. Feeling identification-with-the-aggressor as inner-badness, one dare not take ownership of power. The differentiation of aggressive affects, in a child between 18 and 30 months, but also in developing cultures, leads to envy of those who reap more benefit from the leader, and *schadenfreude* (gloating) toward those who benefit less. The sense of *mine* in primal culture feels like a demonic introject that must be projected outward. That is why American North Coast Indians developed the Potlatch custom of extreme giving—to the point of survival anxiety. Tribal members experience the threat of losing group identity as portending death and decay.

FORMING A CULTURAL DYNAMIC UNCONSCIOUS

Neuropsychoanalytic studies on cultural binding of emotional expression enhance perspective on cultural dynamics. Beginning in young childhood, top-down control over the expression of affect manifests the preoedipal origin of a superego. In archaic culture, taboo prohibits the expression of envy or gloating. The hormone oxytocin increases envy and schadenfreude, emotions that develop in young childhood (Shamay-Tsoory et al., 2009). These affects were regulated by shamans who communicated the leader's rule that greed is profane. As Panksepp says, affects transmit the quality of bottom-up instincts, yet the experience also blends with regulations triggered by cultural prohibitions and others responses. Freud argued that if one person disobeys the leader without punishment by taboo law, the whole group structure can break down.

As king of the psychoanalytic hill, Freud preferred to use his competitive oedipal explanations to preoedipal ones. But by bringing culturally induced guilt back to infancy, he put the oedipal cart before the horse and mixed his oedipal fantasies with his earlier childhood prohibitions. In analyzing adults, Freud realized that obsessive symptoms such as superstition correspond to the role of taboos in structuring cultural laws. Comparing taboo with the prohibition of masturbation in his own culture, Freud showed prohibitions and sanctions in the *touching delirium* of obsessive patients:

> … in very early childhood, the patient shows a strong desire to touch … promptly met by an external prohibition against carrying out that particular kind of touching. … The prohibition is accepted since it finds support from powerful internal forces … and proves stronger than the instinct … the prohibition does not succeed in abolishing the instinct. Its only result is to repress the instinct (the desire to touch). (Freud, 1913, p. 29)

TABOOS ON THE DEAD

If culture is to survive, death must be mitigated in its cultural impact. The dead are taboo. Freud relates evidence of archaic tribes treating the truncated heads of their enemies as kin who need to be propitiated by months of soothing endearments. Warriors were taboo, as if the enemies they killed lurked within them. Thus, killing creates a close relationship. Because rulers can get members of their clan to kill, the rulers are taboo. If a member breaks a taboo, the ruler and his totem exercise even the power of death on the violator. Death as a stage of life is the myth of entering the realm of the dead. Like a river flowing into the ocean, the dead are a source of group identification that religion reserves for the afterlife. Religion cancels death.

Some people wish to cast their legacy of experience and reflection on culture to compensate mourning their self. William Carlos Williams, a New Jersey doctor and poet, who was influenced by Ezra Pound to always write about what's new, remembered to do that after a stroke, when he felt his death was soon inevitable. Preparing to die, not wanting to die miserably, lamenting atom bomb blasts that attacked his hope for progressive culture, depressed about political attacks on him, and regretting his infidelities to Flossie, the love of his life, he wrote *Asphodel, That Greeny Flower* published in *Journey to Love*, containing a famous quote, which I paraphrase psychoanalytically: It is hard to find fresh emotional news, when taboo forces the repression of what is most meaningful in life, until death is inevitable (Williams, 1955).

In *Totem and Taboo* Freud summed up the cultural role of the taboo on the dead:

> The projection of unconscious hostility on to demons in the case of the taboo upon the dead is only a single instance of a number of processes to which the greatest influence must be attributed in the shaping of the primitive mind. … The projection … of internal perceptions is a primitive mechanism … which … plays a very large part in determining the form taken by our external world. … (Freud, 1913, p. 64)

He adds:

> The projection of their own evil impulses into demons is only one portion of a system which constitute the … view of the universe of primitive people, and which we will come to know as "animism". … (pp. 64–65)

Soothing fears of death does more than create a social religious illusion, a myth about nonexistence; it creates a whole virtual world of myth that I think evolves as a form of social phantasy that structures the medial networks of the brain. The DMN is the brain's hub for combining cognitive prediction and fantasies of alternative behaviors into what we may call a virtual social world. Freud was smart enough to realize that speech itself produced a version of reality that could easily become a lie. In that way, the obsessional neurotic could believe his fantasies of killing or rape deserve punishment.

Freud imagined guilt emerging culturally in response to the oedipal act of sons killing the father in order to take what had been his: his wife—their mother—and power over other men. He cites the talion principle—re*tal*-iation—an eye for an eye, a death for a death, saying Semitic circumcision rites maintained the archaic ceremonial sacrifice to propitiate the aggressive god. Culturally induced guilt binds the aggression used for killing, reinforcing the mind's authority over impulses. If culture is a system of group beliefs modifying instincts through phantasy, identification with the dead—including someone you may have killed—gives rise to the cultural fantasy of eternal life.

ESSAY III: ANIMISM, MAGIC, AND OMNIPOTENT THOUGHTS

Freud continues to elaborate the externalization of animism into the cultural world as a religious projection of young childhood: "Animism is, in its narrower sense, the doctrine of souls, and, in its wider sense, the doctrine of spiritual beings in general" (Freud, 1913, p. 75). Continuing, he writes that every culture's animism forms a religious system:

> They people the world with innumerable spiritual beings both benevolent and malignant ... spirits and demons they regard as the causes of natural phenomena ... primitive people believe that ... individuals are inhabited by similar spirits. These souls ... migrate into other human beings; they are the vehicle of mental activities. ... (p. 76)

Magic and sorcery developed as ways of controlling the flow of spirits. Magic procedures are meant to harm bad spirits and help good ones; they can target names or body parts like someone's hairs or nails in order to affect his or her indwelling spirit. By consuming flesh, the eater takes in the qualities of the spirit inhabiting the eaten animal or person—just as a nursing infant or young child seems to absorb the mother's qualities. Freud says that the similarity between the act and the expected consequences is the basis of magic. Every intentional outcome was once observed as another being's action. This connects to what neuroscientists know: when LB acts are followed in less than 0.5 seconds by RB perception, one experiences cause. Intention in young childhood is magical, since words seem to precede acts. Since magic follows associative thought processes, Freud concluded that young cultures, like young children, and like obsessive superstition, see fantasy acts as causal. Magic imitates the totem animal's harmony with nature by sequencing acts to produce their effects.

MENTAL SYNDROMES ARE FANTASY EQUIVALENTS OF RELIGIOUS MIRACLES

In Freud's case studies, animals often symbolize a mental syndrome. In his "Rat Man" case, *Notes on a Case of Obsessional Neurosis*, Freud sees compulsive acts as a magic compromise between sexual impulses and punishment. He attributes this to the father's control over the child's autoerotic behavior (Freud, 1909, p. 208). Among libidinal pleasures he includes feeling food in the mouth, feeling feces passing through rectal mucosa, and the feeling of touching the genitals. His analysis of obsessive-compulsive patients shows that touching deliria following paternal prohibitions lead to the taboo power of religious prohibitions and to false information about the facts of life and death. The child's fear of punishment for using word magic and staging aggressive tantrums leads to abandonment anxiety, which includes loss of the feeling of being one of the family clan.

Summing up Freud's assessment of how archaic culture overcame feelings of helplessness, the primal group follows the leader in projecting an imitation of how animals survive. Defending against loss, archaic cultures deny death through religious ceremonies that magically evoke spirits of the dead to animate their group life. Freud writes about three stages of cultural world views:

> At the animistic stage men ascribe omnipotence to *themselves*. At the religious stage they transfer it to the gods. ... The scientific view of the universe no longer affords any room for human omnipotence; men have acknowledged their smallness and submitted resignedly to death and to the other necessities of nature. (Freud, 1913, p. 88)

Following Freud we must consider whether cultural myths affect brain evolution.

ESSAY IV: THE RETURN OF TOTEMISM IN CHILDHOOD

Freud finished *Totem and Taboo* as the embattled leader of a tribe of analysts, working together to advance civilization. He marshaled analysts and anthropologists to force the challengers, Jung and Adler out of his psychoanalytic society. Instead of allowing them to depose him as the father-leader in a symbolic parricide, they must leave to lead their own tribes. To bolster his own perspective, he cites Wilhelm Wundt's work on how totemic cultures use religion to advance culture from father/leaders, to heroes, to gods.

Freud extended Frazer's *Totemism and Exogamy* (1910). Frazer surmised that the totem was once a source of food for the clan. Freud learned from him that totemism is a group classification; members find collective identity in the totem. Freud concluded that identifying the father with the totem reanimated the primal crime of killing the father and distributing his flesh to all of the brothers. The primal sacrament distributes blood and flesh to the believers, providing a shared, incorporated identification, as it once did with the totem. Since eating flesh comes later than drinking mother's milk, he wonders if there might be two totemic gods: The hunter teaches men how to hunt and pastoralists show women how to raise plants and animals. In our look at Gilgamesh we saw male and female totem gods: Gilgamesh's mother was a wild cow-goddess; Uruk's totem honored by Ishtar was "the bull of heaven," protecting and inseminating the herd.

Freud realized that cultural totemism developed exogamy to project libidinal drive away from familiar kin to outsiders. Familiarity extends narcissistic feelings of kinship; the marital partner is an outside stranger who may become familiar. Since incest leads to family conflict and raw aggression, exogamy intervenes with taboo prohibitions against incest, killing, and cannibalism. The brain's hemispheric duality and prefrontal generalization of experience suggests that the reflective DMPFC divides a LB quality of narcissistic extension from a RB quality of cultural context.

In exogamy, the tribal totem determined kinship on the mother's side; sometimes the father had a different totem. Marriages among tribal subgroups formed alliances that could follow other totems and their taboos, just as exploring the unknown was necessary to survival for many tribes. Eventually, intertribal alliances among progressive cultures branched out to maintain their advantages. Religious tolerance, multilingual members—traders, artists, and story-tellers—and technological innovations for food production and military might were ingredients for bringing heterogeneous tribes into kingdoms.

GUILT MAGIC: RELIGIOUS RITUALS OF SACRIFICE

Freud picks up from William Robertson Smith's account of the ritual meal which brings everyone who participates into communal bond. Reading Smith's *The Religion of the Semites* (1894) as he began his work on *Totem and Taboo*, Freud felt it was just what the doctor ordered. He incorporated Smith's view that group symbiosis in the sacrificial meal originates in eating the totem god. Freud added that when there is a traitor in the midst, like Judas, oedipal consequences brew. He brings guilt back to the primal brothers who cannibalized the father's murdered flesh. Thus, ritual sacrifice brings shared guilt to the members as part of their ambivalent bond.

I had a patient who cared for her sick husband for too many years. Their adult son still lived with them. The two of them often wished the father dead, he finally died on Thanksgiving. When I saw her six weeks later, she bragged about the free turkey she got at the supermarket that had miraculously fed her and her son for the whole six weeks. It was a miraculous sign that she was in charge. So, totem meals can be enjoyed by more than just the brothers. Freud thought that after patricide, survival becomes so tough that a goddess culture brings the mother-protected youngest son into prominence.

Freud asks: How can the god participate in the ritual totem meal when the totem god was once the victim? He concludes that in the evolution of religious thought, since the father of the primal horde becomes both the totem and the god, and since the oedipal murder of the father by the band of brothers comes first and the goddess culture comes next, there is no discrepancy, only development. He agrees with Frazer that much later, early kings of the Latin Tribe were foreigners who could be sacrificed at particular festivals. Thus, like an early childhood wish, the cultural memory of father sacrifice cannot be extinguished and returns from repression.

Freud is happy to bring religion *almost* completely under the oedipal umbrella: "the beginnings of religion, morals, society and art converge in the Oedipus complex … the nucleus of all neuroses" (Freud, 1913, pp. 156–157).

But religion began in his preoedipal stage; the oedipal stage only expanded the scope of religion. Those who doubted miracles of religion, who had enough narcissism to individuate, could go on to the thoughtful analysis of the physical and emotional world. They could reflect sufficiently on their experience and use logic to better their grip on survival. Freud continued to deny the source of his own nagging guilt, the death of his symbiotic counterpart Julius, corresponding to his symbiotic wish to possess all of his mother's love. He wanted to view all guilt as arising from the oedipal crime, rather than from breaking the bonds of infantile symbiosis. Thus he cites the beginning scene of Goethe's Faust: "What thou hast inherited from thy fathers, acquire it to make it thine."

But Freud's exemplar *Rat man* case for obsessional neurosis, *Notes on a Case of Obsessional Neurosis* (1909) shows his own identification with the patient. The patient's neurosis centered on masturbation fantasies about Gisella Fluss, who was Freud's own masturbation image during Freud's adolescence (Harris, Jay and Harris, Jean, 1984). Like Freud waiting to earn enough money to get married for eight long years, the Rat man had earlier wished his father would die so he could afford to marry his fiancée. That oedipal fantasy gave rise to anal submissive and anal aggressive derivatives that focused on rats eating into his anus or his father's anus: punishment for the fantasized oedipal crime. Freud's own wish to have married Gisella was tied into another wish, to have remained in his birth place where the Fluss family remained. Freud longed to undo his early childhood losses, even to have moved to England with his step-brothers and to have married his niece Pauline. He longed for his lost childhood culture:

> … no generation is able to conceal any of its more important mental processes from its successor. For psycho-analysis has shown us that everyone possesses in his unconscious mental activity an apparatus which enables him to interpret other people's reactions, that is, to undo the distortions which other people have imposed on the expression of their feelings. An unconscious understanding such as this … left behind by the original relation to the father may have made it possible for later generations to take over their heritage of emotion. (Freud, 1913, p. 159)

His last words in *Totem and Taboo* are Goethe's: "In the beginning was the Deed" (p. 161). Chapter 8 will deepen our understanding of Freud's subjective approach to his life as the roots of his individuated narcissism.

Chapter 8

Freud's Self-Specimen

Freud's creative development of psychoanalysis is dramatized in his letters to Fliess. I will use the Standard Edition, Volume I, for documentation (1950 [1892–1899]) unless noting excluded portions added by Jeffrey Masson (1985). Deletions from Freud's letters were an attempt by Anna Freud and Kurt Eissler (in charge of The Freud's Archives), and by others to hide the subjectivity in Freud's work. Freud's immersion in creating a logical science of psychoanalysis left him vulnerable to exposing his sometimes irreconcilable basic ideas, such as the formation of mental syndromes due to the effects of fantasy versus trauma. Chapter 8 gets down to the nuts and bolts of his conversion of his life experience into the tenets of psychoanalysis. He revealed childhood and adult experiences and reflections on them in ongoing letters to Fliess, his muse, as he reconstructed his identity formation as a template for understanding others' lives.

> Today's dream has produced … the strangest disguises …. She was my teacher in sexual matters and scolded me for being clumsy and not being able to do anything … I saw the skull of a small animal … I thought "Pig!" … allusions to my present powerlessness as a therapist. Perhaps this is where an inclination to believe that hysteria is incurable has its start. Besides this, she washed me in reddish water, in which she had previously washed herself … she made me carry off "*zehners*" and give them to her …. Where do all patients get the frightful perverse details which are often as remote from their experience as from their knowledge? (Freud Fliess Letter #70, October 3 & 4, 1897, pp. 261, 263)

Julius was the dead, un-kosher animal. Sigismund was the narcissistic pig who insisted on more than his share, a primal attitude that carries primal guilt. Was bloody diarrhea (dysentery) that spread to Julius from Sigismund, veiled by Monika's menstrual blood or her filthy *zehners?*

The essence of a person's subjective experience gives rise to a lifelong set of reflections that fills networks of the brain with a sense of individuated identity, containing the essence of what it means to be alive from the inside out. Freud gives us a template for understanding how conflict shades into fantasies that alter experience at every stage of life; and how those fantasies are worked out as a guide to one's goals and relationships. We want to think of Freud in his expertise, but he was also a self-specimen who could talk to himself and to the reader out of both sides of his mouth. One side was as logical as he could make it: investigating the relationship between dreaming and waking life that could only be resolved by understanding how the brain becomes mind overnight. Fliess had to deal with Freud's logic and illogic and their transferences to one another.

Freud saw male and female development in terms of oedipal identification with each parent in the formation of the ego and superego. He believed that he had worked through his own bisexual identifications better than others. His negative oedipal complex was strong (maternal identifications); his positive oedipal complex (paternal identifications) was even stronger. He would lure other men who also had symbiotic fixations into his group, and leave them in the lurch when he returned to a more logical framework. Freud's early use of hypnosis in treating patients showed him how he could use suggestion to influence others by eliciting a symbiosis. The leader hypnotist makes himself the objective persona; while the subject becomes an adjunctive presence who follows the leader's reflections and decisions. Such an identity split led Freud to model the primal horde on an individualized father-leader and a symbiotic mother as the source of mutual group identity. The leader used necessary aggression to manage the group's survival. His idealized young mother gave pleasure and relief with the milk of empathy.

FREUD, TOTEMS, AND CHILDHOOD NEUROSIS

Making his self-analysis the template for analyzing others, Freud was always wishing to use his cultural experience and his own fantasies as explanations of what he had gone through in arriving at his own maturity as both an individual and a member of culture. I will explore two more animal referenced cases he chose to prove his developmental theory that convey as much about his development as they do about universal childhood development. The cases of Little Hans, *Analysis of a Phobia in a Five-Year-Old Boy* (1909) *and* The Wolf man, *From the History of an Infantile Neurosis* (1918) show each boy's anxious, totemic kinship with male animals: Herbert Graf's horse phobia and Sergei Pankejeff's wolf phobia, respectively. Freud used these cases to bolster his oedipal theories and so reinforce his value to the psychoanalytic community.

Imagining archaic fathers as carnivorous animal killers, who might be inclined to infanticide, he dramatized oedipal fantasy as a universal childhood development. As every patient was a specimen as he was himself, his narcissistic self-transference was always present. His famous *Irma dream* and his narrative of *Little Hans* were meant to repeat Freud's libidinal development with conceptual authenticity. Chapter 8 reveals the extent to which Emma Eckstein (Irma) and Olga Graf (Little Hans' mother) had to be disguised by Freud to protect them and his local community. He struggled to keep friends safe from notoriety, sometimes hiding their traumas as if they were fantasies—which distorted some of his case histories. We'll also see how he managed their lives with rationalized interventions to add to his accomplishments.

RECAPITULATING FREUD'S EARLY CHILDHOOD

Freud's losses during his passage from infancy to young childhood set him up for narcissistic depression mixed with hysterical and obsessive features. He realized that the source of obsessive doubting was a cognitive capacity beyond one's emotional capacity. He had a talent for recovering visual images from his infancy and young childhood and perceiving them in his dream hallucinations. They were a pathway for him to the brain's transformation of perception and memory into fantasy and dreams as a primary process. It is clear enough from the first section of *The Interpretation of Dreams*, entitled *The Scientific Literature on Dreams,* that Freud used his talent to study how the brain produces mind. In the preliminary section, *Memory in Dreams,* he uses an example from himself to prove that episodic memory is imbedded in the brain's associations. The one-eyed-doctor whom he had forgotten for 37 years appeared in a dream the night before his father's funeral—triggering his self-analysis. His mother confirmed the image of his early childhood doctor to be real.

Emphasizing Freud's infancy, I want to reinforce the proposition that Freud's narcissistic compensation for early childhood losses formed mind, brain, and cultural compensations used to create psychoanalysis, which may have been as rational and irrational as religion in providing his *weltanschauung.* Inventing psychoanalysis with Fliess as his witness made Freud vulnerable to every fantasy and dream he had as possibly the truth of memory. Communicating with Fliess to the extent of making himself an open book made Freud vulnerable to him, but also made it possible to reconstruct his own mental development. I think mind and brain together *reconstruct* and adapt memory of one's identity narrative daily, in order to deal with the present.

Fliess too had a creative mind, willing to be irrational to find new medical ways to extend life. Freud and Fliess almost had a *folie a deux* about merging their theories. As his doctor, Fliess treated Freud for cardiac symptoms and migraines they thought to have origins in Freud's sexual disposition. He used Freud's discovery of cocaine's effects as nasal treatment. In the early days of inventing psychoanalysis, Freud thought that since he felt hysterical maybe his father had sexually seduced him. He and Fliess thought that a passive response to another person's sexual stimulation represents a female sexual aim.

At first, Freud confused oedipal fantasy with seduction/trauma as the source of hysteria. But trying to sort out fantasy from trauma, Freud remembered the pleasure his "clever and ugly" "old" nanny induced in him, concluding that his phallic sensitivity and addiction to masturbation were more likely her doing than his father's. He wondered if the seduction he experienced in infancy and young childhood was Monika's touching his penis to soothe him. Sigismund might also have responded to witnessing his parent's intercourse when he and Julius lived with their parents in a single room above a locksmith's shop.

Writing his self-analysis in letters to Fliess who lived in Berlin, and considering fantasy rather than seduction as the cause of hysteria, Freud mentioned a left-sided migraine. Given Fliess's theory of left-sided degeneracy toward male femininity, Freud scoured his past to prove he was not degenerate. He imagined that if his self-analysis could distinguish hysterical fantasy from seduction/trauma, he could become a professor. With excitement he realized that his dreams' visual images played back realities of early childhood; while his memory of spoken words was a record of fantasy formation. He agreed with Fliess that the LB accounted for verbal fantasy and the RB for visualizing reality. He hoped to parlay these aperçus into professorship.

As the original analyst it was also hard for him to distinguish screen memories covering memories of loss or trauma, from the effect of repression on early childhood sexual fantasy. He thought it was a normal cultural practice for nannies to sooth young children by stroking their penis or clitoris. Many of his male patients followed their nanny experience into fantasies of passive sex with prostitutes. The virgin/whore fantasy had deep roots in Freud's culture. Freud was addicted to his nanny as a mother substitute: "If I succeed in resolving my own hysteria, I shall be grateful to the memory of the old woman who provided me at such an early age with the means for living and going on living. As you can see, my old liking for her is breaking through again" (Freud, 1897, p. 262).

Monika was jailed for stealing his zehners when the family was already planning their diaspora. A dream memory of his mother Amalia's disappearance while pregnant with the next child condensed into Phillip's pun: Monika and Amalia were *boxed up* (jailed vs. confinement before birth), which Freud associated to the empty pantry, the box he fell from as a small boy while reaching up for a sweet in the pantry—before the one-eyed doctor stitched up his chin:

> ... the dream picture was a memory of my taking money from ... my mother. So far was I from knowing that she [Monika] was a thief that I made a wrong interpretation. I also made enquiries about the doctor we had in Freiberg, because a dream showed a great deal of resentment against him .... My mother then told me that the doctor in my childhood had only one eye .... If the old woman disappeared so suddenly, it must be possible to point to the impression this made to me .... My mother was nowhere to be found: I was screaming my head off .... (Letter 71, p. 264)

Freud could not maintain conscious continuity between his kin and his family romance couple of Monika and the doctor. Primal players in the ambivalent dynamics of his cultural unconscious, they held his wishes for libidinal satisfaction in thrall. They existed like distant ancestors in his prehistoric past. Following Goethe's lead about how to reconstruct the archaic past, Freud searched his self-analysis for missing links in mammalian and cultural evolution.

FREUD'S SELF-SPECIMEN DREAM

Irma's Injection, the specimen dream of *The Interpretation of Dreams* expresses Freud's wish to show how dreams are energized by wish fulfillments. His patient Emma Eckstein, Irma in the dream book, was the subject of Freud's and Fliess's therapeutic experiment to restore her female/male balance of biological and chemical sources of sexual need. Her repetitious clitoral masturbation after the death of her husband felt wrong to her, and she sought relief in analysis with Freud. Fliess, who was an ENT surgeon, viewed the left and right nasal turbinates as targets of sexual chemicals. Using cocaine to alter sexual chemistry more than as an anesthetic, he had already operated on Freud's turbinates, and on others, with apparent relief.

If Emma practiced too much male-type masturbation since her husband died, they would restore her balance by excising over-productive chemical cells in her left sinus. In early March, 1895, Fliess operated; but he left a piece of gauze in her sinus that triggered infection and bleeding two weeks later (Masson, 1985, pp. 116–118). Freud recorded his associations to the specimen dream for years after the event. Emma was his wife's friend and Freud had copious associations about her and about his own wishes for recognition. Emma was from a Jewish Viennese family in which women fought for their rights. Many of Freud's cases were family acquaintances, Jewish, liberal, modernistic, and proud of being in the intellectual forefront of culture. Emma's three-year analysis was also a way of studying what Freud knew about women, sexual development, and identity. Freud gave her credentials to become an analyst herself. He was proud she used his theories to later prove their psychological relevance in Emma's cases.

Mixed into his thoughts about cocaine as a source of instinctual drive, he had many associations to his dream theory as he went along. He also had associations to Fliess that he tried to disguise in his writing. Freud wished to exonerate Fliess and hide any self-blame, since he felt his analysis helped Emma. His ambivalence toward Fliess was suppressed because he needed him for his whole project. Yet, I think Freud defied Fliess by presenting the Irma dream as his specimen against Fliess's worried advice.

Freud had another female patient during those years that had as much effect as Emma on Freud's understanding of mental symptoms. In letter #60 to Fliess, April 28, 1897, he described a patient who "proved" his seduction theory: some men rape family members. I infer the letter describes Olga Hönig's childhood experience. She started her analysis with Freud when he formally began self-analysis after his father died. As we shall see later in this chapter, her story fits the mother of Freud's case of Little Hans. She

> had a brother who died insane; and her main symptom (insomnia) first appeared after she had heard the carriage that was taking him to the asylum drive away from the front door. Since then she has suffered from anxiety at driving and a conviction that there would be a carriage accident. Years later, the horses bolted during a drive and she took the opportunity of jumping out of the carriage and breaking her leg …. (p. 246)

She told Freud she had been sexually abused by her father from the ages of 8 to 12. He would excite her genitals, and ejaculate on her legs without penetrating her vagina. He had done the same to her older sister and a cousin. When Freud surmised that the father had done worse to her in her younger childhood, she agreed. But, if she proved his seduction theory of hysteria, Freud wondered if his own hysteria was caused by his father. He was confused about hysteria; his seduction theory found fathers or brothers culpable; but his fantasy theory saw hysteria as innate fantasy, accompanying childhood masturbation. Uniting the theories in his next letter, he said fantasies veiled abuse by blocking memory.

Eight months later in the December 22, 1897 letter, censored in the Standard Edition, but restored in the Masson edition, Freud recounts a patient's traumatic memories corresponding to Olga's. The letter describes an infant watching her mother bleed profusely after the father's brutal rape, then at age two she was raped by her father and infected with gonorrhea, then at age three she watched her mother in agony after rape by the father. The patient's descriptions are so explicit and piteous that Freud is beside himself with wishes to rescue her:

> ... authenticity of infantile trauma is borne out by the following little incident which the patient claims to have observed as a three-year-old child ... . She has good reason for identifying with this mother. The father belongs to the category of men ... for whom bloody injuries are an erotic need. When she was two years old, he brutally deflowered her and infected her with his gonorrhea ... . The mother *now* stands in the room and shouts: "Rotten criminal, what do you want from me?" ... Then she tears the clothes from her body with one hand, while with the other hand she presses them against it .... Afterwards she wrings her hands, sits down in a corner ... with her features distorted with pain she weeps ... . When the girl was six to seven months (!!) old, her mother was lying in bed bleeding nearly to death from an injury inflicted by the father .... (Masson, 1955, pp. 288–289)

Imagine a household in which a father's sadistic terror ruins every life. Repeated trauma ruins the lives of the children. If this is Olga, we can understand that her two brothers were suicides, her older sister attempted suicide—and her daughter eventually committed suicide.

Freud went on in the above letter to show the authenticity of traumatic hysteria:

> ... how does the patient know that in attacks one usually enacts both persons, *self*-injury, *self*-murder, as occurred here ... the woman tears off her clothes with one hand, like the assailant ... with the other holds onto them .... Words, whole clauses and sentences are blacked out ... the rest becomes unintelligible. (Masson, letter # 60 to Fliess, April 28, 1897, p. 289)

Freud knew that blocking out words that accompany the memory of trauma, can hide the trauma from self and others. Depending on family members of his patients to pay the bills and satisfy their conscience about hiding family facts, Freud suppressed his knowledge. But, he was bold enough in making interpretations and reconstructions with his patients.

BREAKING WITH FLIESS WHILE PUBLISHING "THE INTERPRETATION OF DREAMS"

Freud clung to Fliess as he wrote *The Interpretation of Dreams*, but Fliess withdrew, feeling Freud was a "*mind-reader*" who substituted his ideas for others' ideas. Fliess did not want Freud to reveal Fliess's guilt in Irma's treatment. He warned Freud not to destroy their circle of friends by revealing intimate details in dream associations. He wrote that just as Freud's expressed death wishes toward him alienated him, Freud's death wishes would alienate other friends. Freud seemed to give in, but revealed enough in his specimen

Irma dream to indicate Fliess's responsibility for her near death. His symbiosis with Fliess was over.

The problem with Fliess as a revenant of Julius was that for Freud, Fliess was a phantom from the past, which he could only get rid of through logic. We see Freud's split between a symbiotic and a non-relationship with Fliess—death wishes that he revealed to Fliess. Despite their mutual excitement as Fliess responded to each chapter Freud sent him, their relationship lost more and more intimacy. Just as Freud later cut loose from Jung in their race to the bottom of archaic consciousness, as he was finishing *Totem and Taboo*, he cut loose from Fliess's illogic as he was finishing *The Interpretation*. Freud gave Fliess more than enough rope to hang himself, just as he always defeated the illogic of straw men in order to organize his own writing. After the book was published, their letters trailed off into polite disagreements over which of them first wrote that everyone has a bisexual disposition—leaving Darwin out of account.

In *The Interpretation of Dreams*, (1900) leading up to the Specimen Dream, Freud interpolated a postscript meant to put Fliess in his place as no longer part of himself. Rewriting the 1909 postscript in 1911, Fliess is no more than an after-birth. He shows Fliess providing an alternative theory of dream formation based on 23-day male and 28-day female cycles, which feature male or female kinds of memories:

> the subject matter of a dream … is to be explained as an assemblage of all the memories which, on the night on which it is dreamt, complete one of the biological periods, whether for the first or for the nth time. A personal communication from the author led me at first to suppose that he himself no longer took this theory seriously, but it seems that this was a mistaken conclusion on my part. (Freud, 1900, p. 94)

The footnote indicates Freud's noting that although he said in 1909 that Fliess no longer held these views, in 1911 he retracted that. Julius as the revenant in all of Freud's irrational companions was first real and then unreal, but still in his memory. Freud's splitting on that basis allowed him to explore the irrational, the archaic, but then to flip over into intense logic.

MEDDLING WITH OLGA

In 1900 Freud introduced Max Graf, a dazzled member of his Wednesday Evening study group, to Olga. Asked for advice about whether to marry her, Freud told Max that she would be an exciting wife and a good mother to their children. Many years after their divorce, The Freud Archives released Kurt Eissler's interviews with Herbert Graf (Hans of Little Hans) and Max Graf. With that data, Franco Borgogno (2013) wrote extensively about the Olga/

Max relationship: Olga and Max were both musicians, both products of traumatic, sadistic, families, and both acquainted with family suicides. Their son Herbert (Little Hans) born in 1903, grew up to become the stage manager and director of the New York City Metropolitan Opera.

Freud's case history of Little Hans, *A Phobia in a Five-Year-Old Boy* (Freud, 1909) provides the case data I used for my comments in the following seven pages. When Hans developed a horse phobia after his sister was born, godfather Freud offered to supervise Max in educating Little Hans about the facts of life—to cure him. It was Freud's chance to prove that sex education prevents neurosis. It's an odd idea that presenting the facts of life to a child will protect the child from developing neurosis. But Freud felt that discovering the oedipal complex remade his own inner world. Little Hans was to be Freud's exemplar for helping a child resolve his oedipal complex. I think he imagined that his analytic insights had cured Olga.

Under Freud's direction, Max informed Hans that females do not have penis; they are not castrated, and boys will not be castrated for wishing to keep their mothers to themselves; babies grow in the mother's uterus from the father's seeds pushed by the penis inside the mother's vaginal opening—not from the stork. Max was also to tell Hans that he loved him and would not punish him for "coaxing" (cuddling) with his mother; and boys wish to grow babies inside their bodies too; but babies come out of the female body, not from the same opening as "lumfs" that go into the toilet. Hans was a petri dish in the germ-filled milieu of Olga's, Max's, and Freud's conflicts.

Single-minded about helping civilization progress, Freud used Olga's family to further psychoanalysis. He let his aims justify his means as long as he could keep his manipulations half-secret from himself. Freud wanted to establish a world of cultural good. Little Hans' developing libido may have contributed to his horse phobia, but the case write-up left his mother's traumas and repetition compulsions through their identification with the aggressor out of account. Freud did not address Olga's brutality to her children nor her sexual anesthesia, which Max described to Eissler. I think Freud wanted to repair her by managing her life. In a similar way, he trained Emma Eckstein as an analyst as a reward for almost sacrificing her life to loving him. He became fixated to that mode by repeating it with the women he treasured.

FREUD'S PRESCIENT GIFT

When Hans's mother was pregnant with her second child, Hannah, Freud gave Hans a rocking horse for the boy's third birthday, carrying it up four flights to personally deliver it. Think of Olga as the patient whose leg was injured in the over-loaded horse accident. Let's reconsider a quote from Freud's letter #60 to Fliess that ties Olga to his 1897 patient: "She has suffered from anxiety at driving and a conviction that there would be a carriage accident. Years later, the horses bolted during a drive and she took the opportunity of jumping out of the carriage and breaking her leg ..." (Freud, 1897, p. 246).

The story of Hans's phobia of horses is rooted in the boy's fear that an overloaded horse will crumple, make a great kick, and injure someone's leg. As his mother's companion, he often listened to her stories of what had caused her emotional pain. Coaxing in bed, even after his sister's birth, would relieve his mother from having to go out into the world, where she felt so anxious. Hans was afraid of losing his mother: if she got hurt she would abandon him. Hans's therapy revealed that in his fear of being bitten by a horse something black around a horse's mouth frightened him. Following Freud's advice about learning the sexual truth, Hans was allowed to view his mother urinating and defecating. Perhaps the black did not represent Olga's feces as Max thought, but pubic hair framing a fantasized vagina dentata.

Freud and Max Graf present evidence that oedipal fantasy leading to castration anxiety was a source of Hans's phobia. But their educational input was just more fodder for Hans's fantasy life of positive and negative oedipal wishes. Hans's identification with his mother and his concern for her hurt feelings is as strong as his identification with, and fear of, his father. Most telling was his game of reliving an accident he witnessed: A friend was hurt and bleeding. In his fantasy, Hans was an overloaded horse who falls, kicking out so hard that his friend's foot was injured. Hans repeated that fantasy/trauma over and over with great affect: "*Cos of the horse! Cos of the horse! Cos of the horse!*" Hans invented stories to tell his father: Horse fantasies were dangerous games that seemed real. In Hans' identification with his totem, an injured horse injures.

Hans also showed phobia toward giraffes on a trip to an animal park. I imagine the giraffe picture in his new room after his sister was born were to comfort him—like the rocking horse. Hans often said his mother had a white neck like a giraffe, canceling the black around the horse's mouth. He told his mother a fantasy while coaxing with her: "In the night there was a big giraffe in the room and a crumpled one; and the big one called out because I took the crumpled one away from it. Then it stopped calling out; then I sat down on the crumpled one" (Freud, 1909, p. 37).

Since that followed Hans having been told his mother had no penis—like his sister, Freud and Max took it as confirmation that the big (phallic) giraffe was his father, the little one his mother. Given the interpretation, Hans laughed, joking that he wished for a gun to shoot someone dead. The crumpled one is his sister who he replaced in bed after his mother beat her. As I see it, Freud did not want to recognize a fantasy of wishing to get rid of an unwanted sibling. Freud and Max disregarded Hans's identification with his mother, who was now saddled with an unwanted daughter. Despite Freud's okay, both parents had had misgivings about bringing a second child into a marriage they sensed was breaking apart.

For Freud, the birth of Julius was an omen of his mother's disappearance. I think Freud was desperately revising his own history by putting his understanding of his own libidinal development in place of Hans's. Instead of the boy becoming sadistic to his mother to punish her for having another child, Freud wanted Olga and Hans to maintain their *coaxing* love. But Hans said repeatedly that Olga smacked Hannah's bottom very hard when she screamed too much. Olga, the victim, became the aggressor.

SPEAKING OF TEMPLATES

> I have … been thinking with pain of the way in which the adherents of "the normal person" will fall upon poor little Hans as soon as they are told that he can … be shown to have had a hereditary taint. His beautiful mother fell ill with a neurosis as a result of a conflict during her girlhood. I was able to be of assistance to her at the time, and this had in fact been the beginning of my connection with Hans's parents. (Freud, 1909, pp. 141–142)

The above statement is part of Freud's using Hans as a self-object who proves that early childhood masturbation is not a sign of hereditary degeneracy. Hans's preoccupation with touching his penis while fantasizing led to Freud's interpretation that castration anxiety was the source of Hans's phobia. Freud wanted to believe that the wish to kill the father and have his mother to himself was a universal hereditary experience passed down from archaic culture—not a sign of degeneracy. Masturbation was a bugaboo for Freud: what he called "the primary addiction." Sometimes a cigar is not a cigar, but a weapon.

> Alfred Adler … recently developed the view that anxiety arises from the suppression of … the "aggressive instinct"… he ascribes to that instinct the chief part in human events, "in real life and in the neuroses"… Hans aggressive propensities (the hostile ones against his father and the sadistic ones against his mother) … seem to have produced a most striking confirmation of Adler's view. (Freud, 1909, p. 140)

A member of the Wednesday Psychoanalytic study group along with Max Graf, Adler wrote about men trying to prove their masculinity despite feeling lame or degenerate. But in this footnote to the Little Hans case added in 1923, Freud distinguished Adler's analytic understanding of aggression from his own. Years later, Olga and Max went to Adler for counseling before divorce—after Olga decided that Freud had ruined her life.

Kurt Eissler's interviews with Herbert and Max Graf in the 1950s show that Olga exhibited behavior many would call borderline (Borgogno, 2013). Borderline patients often report being emotionally and/or physically traumatized in early childhood. Max told Eissler that Olga was often hysterically upset after intercourse. From a perspective of trauma, being sexually abused in early childhood can cause emotional reexperience lacking words to accompany the body's distress. Projective identification—displacing internalized aggression into others—is a way of escaping unbearable fear.

If Olga had been raped in early childhood, seduced in her latency, witnessed her mother's beating, and had been beaten by her mother; if her sister too had gone through trauma, seduction, and fixation; if the brothers too had been targets of the father's abuse and aggression; then cuddling with little Hans would have seemed like a way to master her trauma. If a traumatized mother endures a sadistic husband's rape, as I think Olga's did, she is further trapped by having a daughter. Once, after inviting Hans to cuddle, Olga threatened him with castration by a doctor when she found him touching his penis. Then Hans put a toy knife into a doll his mother gave him, point-down, splitting the legs of the doll apart. How much of Hans's symptoms of restless crumpling legs could have been identification with his mother's symptoms? Hidden behind opera scenes, Little Hans became Herbert Graf who staged operatic emotions (like his music-minded parents).

I conclude that Freud was mistaken in using Little Hans as a specimen case of childhood neurosis triggered by castration anxiety and a positive oedipal conflict. Olga was too much left out of account. Freud even knew and wrote that it is usually the mother who threatens the boy with castration "when father comes home." Trauma causes doubt about the reality of experience, with detachment and spotty memory. Conveying doubt when trying to tell about it, the victim is often misunderstood. Freud did not understand the neural, emotional, and mental changes induced by trauma. His own situation was not traumatic, but enduring loss of kin. With missing templates for building identifications, Freud would feel a need to substitute archaic imagoes to mediate his relationships.

FREUD'S PRIMAL ANALYTIC SCENE

We left Jung behind in the aftermath of Gilgamesh, as he and Freud vied to be the psychoanalytic leader who best understood archaic cultural development. To justify removing Jung and Alfred Adler from leadership in his International Psychoanalytic Association, Freud wanted to produce a seminal work that would encompass childhood libidinal development and cultural development in comparable stages. Jung tried to divide sexuality from libidinal theory; and

Adler tried to use the concept of masculine protest to explain feelings of male inadequacy, rather than considering innate bisexuality. In his *History of the Psychoanalytic Movement* (1914) Freud quoted Adler's protestations to justify excommunicating him: "*Do you think it gives me such great pleasure to stand in your shadow my whole life long*?" Freud said Adler's "will to power" emphasized the instinct of aggression to the exclusion of love: "… he positively considers that the strongest motive force in the sexual act is the man's intention of showing himself master of the woman—of being 'on top'" (Freud, 1914, p. 53).

Once Freud's International Psychoanalytic Association was created, cultural aggressive forces were deployed to maintain its hegemony. Freud used his power to counteract the usurpers: a coalition of brothers symbolically killing the father to take his spoils. The *Wolfman Case* about the wealthy Russian, Sergei Pankejeff (SP), is Freud's tour de force answer to libidinal development, deployed in his characteristically creative way (1918 [1914]). SP was rich enough and treated long enough to feed Freud and his family to the brink of World War I. SP was not only a better vehicle than Little Hans for proving libido theory and the value of psychoanalysis, but another self-object for reconstructing Freud's own libidinal stages. Freud's analysis of development presents a *universal* primal scene passage out of infancy, commodious enough to include Freud and SP. His term *universal* means everyone's and every culture's development. It shows how archaic evolution brought cultural infancy to an end, setting up a later stage for universal oedipal passage in human development.

On Christmas morning, his fourth birthday, the boy dreamed a fulfillment of his wishes to receive double presents from his parents. He dreamed of wolves standing erect in a walnut tree. Using SP's associations to the dream during four years of analytic work, like using images in a myth to reconstruct past history, Freud and SP reconstructed this scenario: SP awoke from malaria fever in his parent's bedroom late one afternoon in his 18th month to see his father standing in a wolf's erect position, making animal sounds, repeatedly inserting his penis into his kneeling mother's buttocks. SP saw his mother had no external genitals. It gradually became clear that the dream veiled witnessing his parents' intercourse when he was 18 months old. He wished to take his mother's place in that primal scene, but that wish recoiled into castration anxiety.

SP began analysis at age 18 after contracting gonorrhea and feeling that his body was going to pieces. Castration anxiety followed the wish to be in his mother's place. His neurosis bloomed in defense against the wish for penetration by his father: he could not defecate except with an enema. Defecation meant expelling a baby sired by his father. Freud educes this with many quotes from the patient: "The world, he said, was hidden from him by a veil …. The veil was torn, strange to say, in one situation only … when, as a result of an enema, he passed a motion through his anus" (Freud, 1918, p. 99).

SP was fixated to the negative oedipal stage where identification with his mother in order to be in her place with his father, meant sacrificing his maleness and his body's integrity.

Underlying the negative oedipal stage was the primal scene—underlying the primal scene was the wish to be back in the womb. That wish contained his identity myth: "just before taking leave of the treatment … he remembered having been told that he was born with a caul … the veil which hid him from the world and hid the world from him. … wishful fantasy … exhibited him as back … in the womb" (Freud, 1918, p. 99).

The case history is full of convincing detail of the oral and anal stages that SP went through in organizing his severe neurosis. Freud explored whether the intensity of SP's symptoms transferred back to the past, or whether the pasts' intensity colored his present. Was Jung right in thinking that a primal scene gets superimposed on the past, not as a memory, but as a fantasy of rebirth? Could fresh associations create a fantasy past? "All the alleviations which the theories of Adler and Jung seek to afford us come to grief … it seems to me more probable that the fantasy of re-birth was a derivative of the primal scene than … a reflection of the fantasy of re-birth" (Freud, 1918, p. 103).

In Freud's mind archaic inheritance overdetermines a person's developmental passage:

> … scenes of observing parental intercourse, of being seduced in childhood, and of being threatened with castration are unquestionably an inherited endowment, a phylogenetic heritage; but they may just as easily be acquired by personal experience … a child … fills in the gaps in individual truth with prehistoric truth; he replaces occurrences in his own life by occurrences in the life of his ancestors. I fully agree with Jung in recognizing the existence of this phylogenetic heritage. (Freud, 1918, p. 97)

Screen memories hide younger wishes. Freud identified with SP's primal identity myth of a caul with which they were both born. Freud's foretold his greatness—confirmed by a fortune teller seer (like Monika); while SP became a great patient for several analysts: If SP returned to his mother's womb he could contact his father's penis. Such was Freud's narrative of the cultural unconscious regressing back through childhood.

THE COEVOLUTION OF HUMAN CULTURE WITH THE HUMAN BRAIN

Freud's position is that we share innate readiness to experience endogenous trauma that shatters our libidinal expectations and promotes a new stage of development. From my perspective, endogenous trauma is a dual-hemisphere alternation of passing from libidinal to aggressive stages of development in the brain's maturation (Harris, 1998):

1. The infant's primary process adjusts to reality with massive cortical reorganization.
2. The young child's corpus-callosum matures to match LB language with RB reality.
3. A top-down reorganization of prefrontal executive functions creates revised identity.

Each stage of libidinal maturation, I believe, is followed by a stage of cultural aggression that inhibits the wishful qualities of the libidinal stage.

Is aggression an instinct of cultural control? Infant hatred, young child envy, oedipal competition, and latency sadism: these childhood manifestations of aggression are enacted, displaced from original targets, or fuel syndromes. And every syndrome includes cultural prohibitions through suppression, inhibition, or indwelling repression. Freud's theory of neurosis saw libidinal wishes as the impetus for pleasure, with cultural power as the moral force for controlling behavior dangerous to the group.

TOTEM ANIMALS AS NATURAL KIN

The image of the wolf portrayed in Little Red Riding Hood, the fairy tale book SP's sister showed him whenever she wanted the thrill of torturing him, brought out SP's wolf phobia every time. Yet the boy still liked wolves and sheep dogs. Animals are innate targets of ambivalence and empathy. Children have no pretension of superiority over animals. Observing babies learning to talk, I noticed their affinity for animals and other babies: an innate narcissistic kinship. In *A Difficulty in Psychoanalysis* Freud wrote:

> At the level of Totemism primitive man had no repugnance to tracing his descent from an animal ancestor. In myths ... the gods take animal shapes ... in the art of earliest times they are portrayed with animal heads. A child can see no difference between his own nature and that of animals. He is not astonished at animals thinking and talking in fairy-tales .... Not until he is grown up does he become so far estranged from animals as to use their names in vilification of human beings. (Freud, 1917, p. 140)

Pursuing the totem connection of human descent, Freud asked other analysts for write-ups of their cases of animal phobias in children. Ferenczi reported his case to Freud in their correspondence while Freud was writing Totem and Taboo. Ferenczi (1994) reported Árpád, the "rooster-boy" who at two-and-one-half was pecked on his penis by a rooster when he was peeing in the hen house. He had an injury with bleeding and needed a bandage. To pacify him, the bird was killed for supper. Thereafter he was fascinated to watch intercourse of cocks and hens; he imitated *cock-a-doodle-doo* so completely that he awoke the household every morning for the next two years. Instead of speaking, he imitated the sounds and actions of his animal totems. Dancing with excitement, nothing gave him greater pleasure than to see a chicken killed, or to kill it himself. Unfortunately, in identifying with the cock that pecked him, torturing women became the boy's perverse response to his castration anxiety. Ferenczi understood Freud's concept of regression from the oedipal stage to anal-eroticism and anal-sadism: Arpad's identification with the aggressor imitated the totem-animal, by taking the part of the sadistic aggressor. Young childhood word and sound magic, using imitation to find the power of aggression, restructures oedipal age fantasies.

This Chapter has dealt with Freud's problem of distinguishing syndrome etiology. In neurosis based on oedipal fantasies, interpretation can resolve symptoms of conflict. But when trauma and loss are syndromatic sources, reconstruction is a better therapy mode. Detachment is the natural way of dealing with unbearable emotional and physical pain. In a detached state, defenses against regressing to earlier states of ego identity are set aside. Since memory is unreliable in regression, reconstruction, for example generalizing the experience, is more effective than interpretation because it relieves remembering events and episodes. Reconstruction replaces missing judgment and self-observation, which is manifest in the detachment of trauma and unbearable loss.

Later chapters will consider how the neural mechanism for producing fantasies of change is used in the formation of myths, which are cultural ways of denying collective trauma. Phantasy transforms the unwritable history of past catastrophe. Freud's ideas of culture came from a reconstruction of his own preoedipal archaic life, put to the test of psychoanalytic logic. In chapter 9 we will see how the construction of his narcissism through identification with Goethe led to his use of subjectivity to feel his way into the essence of his creativity. I think that despite Freud's creative revision of his patient's lives to make them conform to his own defenses against his preoedipal losses, his theory still sets our valuable truth about the dynamics of individual and cultural development.

Chapter 9

Freud as Goethe

One can identify with another person's objective qualities or wish to be that person. Chapter 9 reveals how much Freud followed Goethe as his cultural leader by giving himself the right to absorb Goethe's creative narcissism. His identification with Goethe was subjective: what he inherited from that forbear father, he made his own. Like Goethe, Freud used imposture to impact culture with his fantasies. He applied Goethe's explicit lessons to his own era to understand bisexuality, fetishism, the effect of sexual and narcissistic tendencies on cultural leadership, and how scientific technology leads to cultural progress. Freud made Goethe the template for conducting his own cultural life, expecting psychoanalysis to provide the same template for others to live their lives by.

> An unconscious understanding ... of all the customs, ceremonies and dogmas left behind by the original relation to the father ... made it possible for later generations to take over their heritage of emotion. (Freud, *Totem and Taboo*, 1913, p. 159)

This quote refers to cultural practices reinforcing cultural continuity over generations, showing that cultural evolution makes a cultural unconscious regulating a biological unconscious. Freud looked to builders of civilization as extended kin; Johann Wolfgang von Goethe was preeminent among them. Freud could not write a paper without quoting Goethe, his exemplar of a leader who altered his surrounding culture to enhance civilization, while pursuing his *Bildungsroman* as the hero of his own novel. Freud used that technique to create psychoanalysis. His identification with Goethe, his wish to be his generation's Goethe, explains how Freud built his theories on Goethe's reflections as much as on his own experience and his own father's experience.

The narcissistic beauty of men working together to enhance culture shows up in a base form. One of my homosexual patients bragged about the beauty he and two friends produced, masturbating in an alley together as they all came together "like a fountain." The literary magic of masturbation shows up in *Faust*, Part 2. Goethe tells the story of a homunculus who creates a wife for an explosive short life together. In his *Autobiography*: *Truth and Fiction,* written after he was sixty, Goethe posed as the omnipotent narrator for his beloved younger self. Exquisitely recalled childhood experience filters through his reflection, like a marionette play of past culture (1848). His reflections on his own childhood showed Freud more about libidinal development than any other case history. Goethe provided pervasive perspective, an array of sophisticated ideas for Freud's writing on culture. As a child Goethe even learned Hebrew, getting an expert to tutor him, so they could discuss the Old Testament. As a child, he concluded (as Freud did as an old man) that Moses was an Egyptian killed by his followers, who had become wandering Jews.

Goethe's autobiography gave an eye-witness account of the customs of the Holy Roman Empire followed in Frankfurt. Even as a child he knew that ceremonies, rooted in centuries of history, bound aggression that could tear the Empire apart. Freud's German sense of cooperative culture among many Austro-Hungarian Empire tribes and clans was rooted in Goethe's elaborate descriptions. But Goethe was also a herald of modernism. *Faust*, Part 2, laments technology that made land suitable for larger populations, leaving the old people with their linden trees forever without a familiar place to call their home. He showed Freud that technology leaves customs of tribal and family life behind.

AUTOBIOGRAPHY AND MEMOIR: NARRATIVES OF CULTURE

Latency is when children absorb cultural myths about their identity sources. A precocious Freud read Goethe, Shakespeare, Cervantes, and much more of the world's literature. He never stopped studying how the greats effect social organization. He learned that they developed a lifelong guiding motif to produce a cultural impact consistent with their developing identity. Using the impetus of their narcissism, they pressured society to conform to that identity. Tracing his Jewish ancestry to Cologne in Roman times in order to feel his kinship with Goethe, Freud felt like a member of progressive culture.

An autobiography like Goethe's, or Freud's throughout his work, is as close to fiction as to truth. Both absorbed Jean Jacques Rousseau's *Confessions,* a prototypical autobiography of truth and fiction, written to impact culture. Rousseau's impact on the French Revolution was characterized by torturous confessions of emotional wrongdoing. Like Goethe and Freud, he had guilt about his fictions—attempts to impact culture with ideals he only pretended to follow—for example, using fiction to compensate for his own childhood losses. He presented ideals of French pre-Revolutionary women mothering their children with breast milk, while sending his own children to catholic adoption after their birth, leaving them in unknown hands: dead or alive. If Goethe and Freud were members of the elite, Rousseau was a member of nature's tribal beginnings.

Both Goethe and Freud presented enough supposedly hidden data about other's secrets to nullify their loyalty in relationships. They treated others as exemplars of what they had practiced in their conceits—as if others' lives were fictional, which gave their writing a literary relish. *The Interpretation of Dreams* is Freud's memoir, encoding his dynamic cultural unconscious, which eventually led to the Goethe Prize for his writing style. Mature narcissists place hierarchical cultural development above any relationship. Culture uses the aggression of narcissists to insure its control over ordinary individuals. Freud's psychobiography is the lodestar of dynamic psychotherapy. He realized that a first memory records the birth of the cultural self—to be preserved at all costs, as personal mythology. Authors and other narcissists turn their reflections on how they survived into fictional characters that impact culture.

SIGNS OF LOSS AND TRAUMA

How Culture Runs the Brain investigates how trauma and/or unbearable loss enter individual and collective cultural phantasies as restitutions for loss, and revenge for trauma. Attempts at developing minds that can survive great stress pervade eras of cultural change. Freud contributed to understanding how trauma left clinical signs that distinguish fantasy-based neurotic symptoms from symptoms that signify trauma. For instance, he learned from his patients that if a person has no memory of latency, that points to a highly disturbed childhood. I had a 26-year-old patient who wanted to become a policeman. He had been abusively beaten by his father during his latency. When he saw a black man attacked in the street by the police, others were frightened, but he casually walked away. The next morning he stayed in bed with his teddy bear. His detached identification with the aggressor was less strong than his wish for mother.

Trauma detaches a person from experience into reflection, but grief stimulates fantasies of recourse. Loss of a loved one leaves a void that triggers memories that may feel unbearable. Children cannot grieve; until adult life they cannot know what has been lost. Until then, others are templates for constructing ego identity. Children do need to imagine new culture to replace what is missing. Religions may help grieving adults by enacting ceremonies that portray myths of eternal life within the culture itself. Goethe and Freud and other shamans can teach therapists how to deal with loss and trauma. I treated a woman who had significant childhood losses as well as later sexual trauma. She became a famous singer who only felt really alive and whole in her identity when she sang her own creative songs for her audience, pouring out her affect so that others could absorb her lessons. Like religion, fantasy thinking compensates for loss and trauma.

ABSORBING ANOTHER PERSON'S UNCONSCIOUS DYNAMICS

Freud wrote about Goethe's first memory which he read in Goethe's *Autobiography: Truth and Fiction* in which Goethe artfully applied retrospective screening (Freud, 1917). Goethe tells a memory of throwing crockery out the window after his younger brother Hermann Jacob was born, while three grown-up neighbor brothers watch in *delight*, as little Wolfgang switched from throwing *toy* crockery, to *family* crockery into the street. That part of the scenario fulfills the wish that Goethe would delight the world through his activities and have extraordinary success. Unlike three siblings who died in childhood, including Hermann Jacob who died at age seven, those grown brothers were his ersatz audience.

In my reading, Hermann Jacob intruded on Wolfgang's symbiotic relationship with his sister Cornelia, just 14 months younger than he was. I think Freud wished to turn Goethe into another case history of himself. He identified with what he assumed was Goethe's childhood ambivalence about having another boy intruding into his unique place as his mother's favorite. But there was a difference: Wolfgang still had a symbiotic relationship with a living girl sibling in infancy. Just as Freud kept Julius alive in symbiotic memory, building an imaginary relationship with his disappeared double, so Goethe kept Cornelia always alive as his double.

After their father's mother gave them a marionette theatre before she died, Wolfgang and Cornelia created social theatre together when they were nine and eight. With an invited audience of other children and nursemaids, they compensated for losing her—and three siblings who died in their early childhood by creating their own world. Fantasies of a created cultural world enacted in play give rise to a narcissistic sense of culture as self-extension, which compensates for losses of kin. Goethe was a poet of desire, not fulfillment. Instead of loving, he had a talent for cultural synthesis, creating total identifications as he had with Cornelia. In Goethe's *The Sorrows of Young Werther*, which triggered a cultural epidemic of suicide, he worked through relinquishing Cornelia to another man. After she died giving birth at 28; he could not love another woman.

Thus, in his paper on Goethe's first memory based on Goethe's autobiography, *A Childhood Recollection from Dichtung Und Wahrheit*, translated as Truth and Fiction (Freud, 1917), Freud overlooked Goethe's twinship feelings. Goethe always spoke of his childhood with Cornelia and Hermann Jacob as a collective *we*. The wish to possess the parents exclusively can hide earlier feelings of loss. I think that was the case for Freud. In 1919, he added a footnote about Goethe to his 1910 Da Vinci paper: about first memory as a screen: "... throwing out of the crockery ... as a magical act directed against a troublesome intruder ... is to triumph over the fact that a second son was not in the long run permitted to disturb Goethe's close relation with his mother" (Freud, 1910, p. 85).

I think Freud was wrong; although Julius could not totally prevent Sigismund from being his mother's favorite, Cornelia cannot be thrown out the window of Wolfgang's self; she shared his subjective center. There is no total separation from a sibling so close in age; self and other are not differentiated sufficiently to form rivalry within a still nonfunctional reflective cortex. Twinned with Cornelia, Wolfgang developed strong male *and* female oedipal fantasies. His bisexual fantasy produced a strong sense of imposture that appealed to Freud.

Story-telling begins in latency, after a person learns how to lie. In fact, fantasies give birth to lies as a way to make dreams come true. In *Truth and Fiction* Goethe taught Freud how creative writing builds on the base of earlier childhood fantasy. He recounted his legend told to pals so often it became a talisman. It fixed in his memory during the Seven Years' War (1756–1763), a time when reality and fantasy blurred as they do in wartime for latency boys. Wars are times of trauma, loss, and cultural change, when heroic fantasies come naturally to the young. Those predisposed to fantasy thinking use knowledge of their own pleasurable biological processes to explore the insides of their mind. Like Freud's dreams, Goethe's fantasies of his cultural era are materials he used to enter his group's collective cultural unconscious. Goethe's fantasies were so useful to Freud's libidinal theory we'll take a long look, as if we were Freud learning from his master. Young Goethe tells his friends his masturbation fantasy as a legend to impress them.

"THE NEW PARIS: A BOY'S LEGEND"

Nine-year-old Wolfgang, lying on his legend-spinning bed, imagines a holiday costume for Whitsunday services: in a suit cut from his father's golden bridal waistcoat, he is a dandy, hat in hand and sword at side. But anxiety enters his reverie as the suit falls from his body as if by its own accord: Mercury brings a mission from the gods. He is to decide *The Judgment of Paris*: Which lady, dressed in red, green, or yellow, will win the New Paris beauty contest? A Helen of Troy of his own comes dancing on his fingertips—when he tries to grasp her, she boxes his ears. (He and Cornelia had seen a puppet show the day before in which Faust gained Helen of Troy through his pact with Mephistopheles.)

In Part Two he tells his friends about entering through a *bad wall* into a hidden world. A starling cries out "Paris, Paris," another "Narcissus, Narcissus." An old man insists he relinquish hat and sword, and wear boy's clothes to enter the chamber. Done: he goes leaping inside. The old man checks his exuberance, warning that if he breaks the rules he will be subject to the green leather cords hanging nearby. To penetrate to the *central* garden, he tiptoes over a living lattice of sharp weapons. Inside, three beauties gracefully finger musical instruments: a harp, a guitar, and a lute. Their handmaiden is his *fingertip girl*, who provides a game set of living war-pieces, *Amazonians* for her, *Achillians* for him. Girl and boy toss little agate cannon balls at each other's forces. The play gets out of hand: girl, then boy, hurls cannon balls so hard that war-pieces dismember and reassemble with shows of pain. While they go at it full force the girl screams bloody murder.

He cannot escape the living lattice of lances and halyards. They crescendo in fury, shredding his clothes from his body. A climax seeps like a warm bath over him as he and the girl have the pleasure of destroying each other over and over. The old man returns, threatening to turn him into a marionette: The cords are devices of obedience. But he tells the old man he knows the secret of the place and is protected by knowledge the gods gave him, as long as he keeps his prize—the girl on his fingertips.

THE SHAM IN SHAMANISM

Fantasies have sources in archaic legends. By telling and retelling his story to his friends, the reality and the fiction of his legend assume mythical proportions. Goethe learned to conceal and exhibit his masturbation fantasy as a derivative of possessing Cornelia. As a story-teller, he exerts power over the other boys. Setting the stage for his adolescence, he revealed his identity-organizing masturbation fantasy: His bond with Cornelia overcomes all obstacles to his achieving social power though story-telling. In adolescence, Goethe wrote stories for an older group of adolescents, pretending to be a girl or boy in love so as to play tricks on someone the group wanted to tease. Wolfgang was learning his cultural role—imposturous, shamanistic identity. His *Legend* foreshadows his life of writing.

Like the Iliad, *The Legend* is a battle to control the culture of men and women by controlling the narcissistic self: for example, masturbation. In Goethe's mind his sister and he contend for control of the self. His masturbation imagery embodies his fantasy of entering the female body. Wolfgang hurls tiny cannon balls at a sibling who can give almost as good as she gets. But discharge of his raw sexuality and aggression makes castration seem real. With all out aggression only one can survive. The vagina dentata fantasy in Goethe's legend is as real as masturbation fantasy can make it. Freud realized that male perversion turns castration anxiety into fetishism.

Goethe is a voyeur and exhibitionist with castration anxiety relieved through a fantasy of owning Cornelia's body as well as his own. In adolescence he went swimming nude to dismay the authorities. Goethe takes Freud to school on the development of latency, as well as adolescent sexually organizing identity fantasies. Most latency children cannot enjoy mature sexuality, yet they can imagine a sexual identity for themselves. That reminds me of the joke about the man who tells his analyst he had terrible anxiety dreaming he was in a room with many naked, beautiful women. "So why the anxiety?" asks the analyst. "Because I was one of them," the patient replies.

In his *Legend*, his fetishistic description of the bad wall leading to borders of wooly plants before entering the beauty in the inner garden shows overcompensating denial of a missing penis—like Faust begging Mephistopheles to get him Gretchen's ribbons or garters—witches are phallic women. Contrasting beauty and ugliness in all his descriptions of Cornelia in *Truth and Fiction*—disavowing her missing penis, Wolfgang lets his fingertip girl enjoy his body as if it were hers. Goethe reveals his Mephistophelian capacity to morph into male and female disguises to change the world.

In studying perversity (paraphilia) as a combination of bisexual fantasies and an imaginative way of bringing male and female principles together through male disavowal of castration anxiety, Freud elucidated fetishism, exhibitionism, voyeurism, pedophilia, sodomy, sexual sadism, and sexual masochism: all sharing the same mechanisms. He often said that masturbation is the primary addiction. Perversion is addictive, promoting exceptions to normal cultural understanding of love making anatomy and how it works to provide satisfaction. Freud distinguished perversion from psychosis: In perversion, the ego splits between real and fantasy versions of virtual castration. In psychosis, the disavowal of reality is maintained as a delusion; while in fetishism it is maintained as a fantasy condition that denies the possibility of castration (Freud, 1927). Does Freud's belief that men regulate culture together by sublimating their homosexual bonds enact the perverse fantasy of ruling women because they lack a penis?

FAMILY ROMANCE IN CULTURE

As Freud noted, people who impact culture often have their designs set in latency. In Jean Paul Sartre's autobiography, *The Words*, he described his latency feelings of imposture that accompanied his absorption of famous author's words as he claimed their works for himself (1963). He felt the painful otherness of his identity as he tried to influence the adults he met with his genius. Mature narcissists turn their identity reflection to an imposturous-feeling attempt to influence culture on their own behalf. In Sartre's *Saint Genet*, he showed how Genet, having been called a thief when he took what he wanted in a family party, became an antihero, as the word thief provided him with identity. Taking on the identity of imaginary forbears in latency leads to the same sense of imposture. Family romance latency fantasies manifest the will to break the rules or to make up new ones. Terrorist organizations have realized that imaginative children often make the best terrorists.

The Seven Years' War during Goethe's latency provided him with imaginative directions for his male cultural identity. His maternal grandfather Textor, the *Schultheiss* (bureaucrat-in-chief) in Frankfurt owed his diplomatic power to the Holy Roman Empire; Goethe's father was loyal to Prussia's Frederick the Great. Goethe's male identity split between the diplomatic, bound aggression of grandfather Textor, and the raw aggression of his father. That split shows up in Goethe's famous poem *Erlking*, a name for Frederick the Great.

In the poem, the father holds his son in his arms before him on the saddle, as they return home late in the night from a long distance. The boy complains that the Erlking is luring him away with spoken promises of eternal love of his daughters and his wife. The father quiets him by saying that it is just the sound of the wind. But the boy persists, finally beseeching his father's help as the Erlking is wrenching him away. The father's reassuring words and embrace are to no avail. They arrive home with the boy dead in his arms. Like taboo, the imaginary can take over a person's identity.

Nicholas Boyle's biography of Goethe says the 30-year-old Goethe wrote this poem after riding at night with little Fritz von Stein, then 11, on the saddle in front of him as Goethe held the beloved boy in his arms (Boyle, 1992). Goethe felt comfortable in giving up seeking a wife or lover after Cornelia died; when he joined the Weimar court, he made Fritz's mother, Charlotte von Stein his muse in place of Cornelia. Alliance with his muse, a lady in waiting in the court hierarchy, assured him a favored position. Since Charlotte was exhausted after her husband's death, Goethe became Fritz's tutor and took him into his house to live with him. Goethe kept his siblings alive within his self-love, just as god at the finale of *Faust,* Part 2 brings all the dead new born infants into himself.

Freud often said, a boy's wish to yield to the father is the negative oedipal basis of patriotism—laying down his life for the country. The loving side of Goethe's ambivalent wish to be the only son was his wish to protect and enfold Hermann Jakob and Cornelia. He played father to Hermann Jacob, making notes for his education, just as his father constructed Wolfgang's education. In the poem he tries to protect his "son" Fritz, yet identifies with the Erlking, iconic in his era as Frederick the Great, Wolfgang's hero.

On the sadistic side of sovereign power, Frederick the Great's father taught him a lesson about complete allegiance to the Prussian State, by forcing him to watch the execution of his bosom buddy after Frederick tried to run away with him. Frederick was to put no one before his duties or his father. Later, Frederick fell in love with Voltaire's writing, and began a decade long courtship by letter. Voltaire was a paid courtier at Frederick's court, when Wolfgang was a boy. But Voltaire did not want to give in to Frederick's anal penetration. Leaving the sadomasochistic court after three years, he was then held prisoner in Frankfurt until he relinquished the letters that showed Frederick at his most perverse. That episode may well have contributed to Goethe's maintaining a proper distance from *his* Prince. Sexual fantasies play a role in the formation of policies that determine the outcome of cultural clashes.

TAKING SIDES IN WAR: AN ASPECT OF CULTURAL IDENTITY

For three years of the Seven Years' War, starting when Goethe was 10, his family home in Frankfurt was commandeered by the Lieutenant in charge of civilian pacification and administration for the French. His mother tried to make the house a home to her tenant Lieutenant. A highly cultured count, he took Wolfgang's bedroom for his art collections. Wolfgang became the count's go-between with local artists, whom he knew from visiting them with his father. Goethe's father sided with Frederick the Great, the Lieutenant's enemy; while his maternal grandfather was beholden to the Austrian Emperor who sided with the Lieutenant. Goethe's family was emotionally and politically divided. Multiple affiliations taught Goethe to delay narcissistic satisfaction with diplomacy.

During a battle determining whether the Lieutenant's side or Frederick the Great's would prevail, Goethe's father came close to being imprisoned by openly supporting Frederick. Fortunately, there was someone else Goethe could identify with; an adjutant who managed the Count's narcissistic rages, when he felt his kindness to the family and community unappreciated. From the adjutant, Goethe learned to spin a narrative in service of diplomacy. Diplomacy, balancing social ambivalences, was the essence of the Empire's political legacy right down to the Habsburg Vienna of Freud's youth. In the end, the Lieutenant's love for the boy saved the father.

Later at the Weimar Court, Goethe mediated decorum for the University of Jena, theatre troupes, writers, artists, scientists, mining interests, and shifting military agendas. During the French Revolution, when he accompanied Weimar troops against Napoleon, he experienced the intensity of cultural affiliations. He proposed a cognitive theory based on the image of cannon balls rending the very reality of vision and feeling:

> The Prussians moved so slowly that Goethe … had time to set up an experiment on the psychological effects of cannon-balls … on 20 September 1792, they were repulsed by the famous "revolutionary cannonade." After that, war fed off the Revolution, and the Revolution fed off a successful war. (Norman Davies, 1998, p. 721)

Davies says Goethe saw cannon balls distorting images with light—shades of Cornelia.

FREUD'S CULTURAL IDENTIFICATION WITH GOETHE'S POLITICS

> Francis Joseph I (r. 1848–1916), who described himself as "the last monarch of the old style" ruled over a truly multinational state, where the imperial hymn could be sung in any of seventeen official languages, including Yiddish. (Davies, 1998, p. 829)

Freud believed that diplomacy helped culture progress. Francis Joseph, Emperor of the Austro-Hungarian Empire, sided with Prussia in order to maintain a military advantage. Freud respected the Emperor's strong leadership that also promoted liberal individualism.

Using journals to spread their influence, Vienna's multicultural modernism gave individuals opportunity to impact culture. The world had become more hierarchical as contending colonial powers rationalized a right to possess weaker nation states. Prussia had followed Bismarck's *realpolitik* by giving Bosnia and Herzegovina to the Austro-Hungarian Empire in 1878. But Balkan, Ottoman, and Zionist nationalism challenged the early twentieth-century balance of power.

The remnant of a diplomatic Holy Roman Empire in Austria-Hungary balanced a liberal, progressive coalition of intellectuals, capitalists, and Jews in Freud's Vienna. Middle-class idealism held the center of bureaucracy and government, shaping Freud's political identity. Developing more anti-Semitism, Hungary resented Austria's influence on Prussia. By the 1890s, political idealism was breaking down in both countries, yet with more freedom of dissent than other parts of Europe. Ethnic pressure for autonomy drove LB revolutionary tendencies, while holistic, multiethnic *nationalism* resting on diplomatic aristocracy conserved RB culture. It was a time when Ferenczi's Jewish family supported a campaign among intellectuals to increase Hungary's liberal modernism in the Empire.

Just as Goethe joined the aristocracy of Prince Carl August in Weimar, Freud used the influence of his patients to gain the title of Professor. Freud wanted non-Jewish affiliations to make psychoanalysis a mainstream cultural movement. In Freud's mind based on rumors, Jung linked to Goethe not only as Christian, but also as a great-grandson of Goethe. Jung was granted a family romance forebear. Freud wanted to bond with Jung as a self-extension. He felt shared identity with Jung through Goethe as *identity synchronicity,* a proof of Jung's concept of a *collective unconscious.* During their initial face-to-face meeting lasting for 13 awesome hours, when Jung told Freud that he had memorized *Faust* as a child, Freud *knew* he and Jung were related.

In chapter 10, we will extend the concept of Goethe's romantic influence on culture and on Freud. Goethe was a RB thinker who conceived visually that a sheep's skull in a Jewish cemetery that he perused had the same maxillary bone as a human skull. If both skulls were extensions of vertebrates, humans were extensions of mammals. He linked his finding to his poetic imagination, making concepts from holistic perceptions. Like Goethe, Freud too used RB holistic thinking to make scientific conclusions; while his LB idealism informed his romanticism as it had informed Goethe's and Darwin's.

Chapter 10

Modernism and Cultural Disciplines

A psychiatrist and English Professor, Iain McGilchrist applied Freud's theories to cultural sequences in Western history, describing a LB/ RB alternation in cultural eras. LB enlightenment was followed by RB reformation; RB romanticism was followed by LB modernism. He sees LB identity as individualized logic subject to the ambiguity of words, and RB identity as holistic, culturally based, collective identity. The metaphor in *The Master and His Emissary: The Divided Brain and the Making of the Western World* shows *Master* as RB cultural identity and *Emissary* as LB individual identity. Dealing with reality, the master depends on individual emissaries to report back on unexpected details; some emissaries are so carried away with their own ideas that they try to take over.

When leaders rationalize their selfish pleasure and unmitigated power the world becomes ruled by immature narcissists, who impose their will on the rest of humanity. "... it is as if the left hemisphere, which creates a sort of self-reflexive virtual world, has blocked off the available exits, the ways out of the hall of mirrors, into a reality which the right hemisphere could enable us to understand" (McGilchrist, 2010, p. 6). He adds on page 8: "Freud ... believed that the mental entities that he described, and whose conflicts shaped our world—the id, the ego and the superego—would one day be more precisely identified with structures within the brain."

Cultural aggression and control unbinds when there is too much dissonance between leaders and the led. For instance, before the French Revolution, as aristocrats became narcissistic leaders of the masses, keeping too many resources for themselves, the people were deprived of their needs: "Let them eat cake!" Narcissists can become so addicted to their words that they write them over and over: lost like Narcissus in a Versailles hall of mirrors. The vortex starts at the top, collecting true leaders in the down spout, picking up momentum to sweep workers, farmers, tribes, and religious believers down the drain. The French Revolution unbound so much raw aggression that when rumor struck, the guillotine never dried. McGilchrist's metaphor is apt for the present world.

LEFT AND RIGHT BRAIN CULTURAL TENDENCIES

Let's briefly review some of Freud's understanding of how the brain operates when cultural issues are at stake. He believed that language allowed people to lie and manipulate for personal gain. Freud also concluded that verbal memory can be changed by others' statements of what happened; while visual experience remains closer to the truth. But history uses the written word for a sense of coherence. Since the LB sequences data in searching for cause and prediction, words imply a belief in causal process. The role of the hippocampus in maintaining individual integrity of planning and prediction is storing LB declarative memory in before/after sequences of planned experience.

Recording history in written words also makes a culture vulnerable to intervention, since word logic can be manipulated. But RB hippocampal memory maintains affect lessons stored in episodes within the milieu of a timeless emotional life. In terms of cultural messages, LB leaders use word logic, gestures, and command to carry out cognitively planned necessities; but shamans have always dealt with the emotional life of culture. They demonstrate how to follow the leader by their stories of custom and tradition united by group memory. Singing and prosody are the RB part of affective messages in language. McGilchrist says that the affect music of shamanistic stories represents a RB contribution to cultural phantasy (mythology).

The more we can understand about the brain's modes of thought, the more we can understand how that translates to collective, cultural feedback. We have considered three modes of individual thought: practical task-oriented; fantasy-based emotional assessment, and traumatic identity change. At a cultural level that translates to policy, institutional conflict, and cultural evolution or revolution. When individuals or cultures are threatened by their lack of prediction, fantasy about identity changes takes over. The switch to default mode network is frequent during complex cognitive tasks that require many reevaluations: "We propose that the DMN may be recruited whenever large changes of cognitive control are required" (Crittenden, Mitchell, and Duncan, 2015). In *Mapping the Self in the Brain's Default Mode Network*, Davey et al. (2016) come to the same conclusion: holistic cultural identity is at stake.

In stressful emotional crises many people undertake prolonged fantasy searches for pleasure in culturally accessible ways, which often requires changes in individual identity that correspond to cultural niches. Freud characterized people by their sexual (individual) and ambition-oriented (cultural) fantasies. Ambition is a narcissistic force for cultural influence. Cultural identity tendencies toward democracy or dictatorship are a window into each person's ambivalent conflicts. "Whose side are you on?"

In Freud's Vienna, opportunity to participate in Germanic culture pushed ambitious Jews, even whole families, toward conversion. Leaving Jewish tradition behind was unbearable for some, even though identity tendencies in late nineteenth-century Europe were shifting from religious affiliation toward the nuclear family and individuation within a more global economic context. Indeed the modernism of individuality, as opposed to extended family religious practice, resulted in high rates of northern Europe suicides, especially high among Freud's Jewish followers. Suicide not only snuffs out individual life, but also attacks the whole family tree: Wars and revolutions destroy the forests.

Freud had one foot in Goethe's romantic idealism and the other in modernizing brain science. By the time Freud was growing up in Vienna, alliances within the Hapsburg Empire still remained, while colonial military and economic empires were contending for global control. The French had lost to Bismarck's Prussian forces after their 1848 aborted revolution, while Bismarck's Prussian hegemony took over. By 1898 when he died, Bismarck's legacy of selective wars for Prussian dominance was ending. The French were regaining their old glory after their defeat by Bismarck, and Austro-Hungary was still managing to maintain its hegemony. As an Empire it was already on its last RB diplomatic legs. Bismarck felt that the Austro-Hungarian Empire was too liberal. His military had been strong enough to resist Austria's influence, but with so many currents at play—the aristocracy, pan-Germanism, liberals, conservatives, and Catholic political powers vying for advantage in the economic and political spheres—he had to endure a balance of the old and new identities.

THE CULTURAL ECONOMICS OF FIN-DE-SIÉCLE VIENNA: CARL MENGER

Given the Hapsburgs still reigning in Freud's modernist Vienna, I want to describe the effect that Carl Menger, an innovative economist, had on Rudolph the Crown Prince. Modernism may be seen as an offshoot of cultural colonialism by large Western nations using technology and economic innovation to divide up the world. Liberals could influence powerful individuals by redefining the social sciences of economics, psychology, politics, and history in their own rights. Economics in particular was a social science that gave individuals an opportunity to show their knowledge of the economics of hierarchical world power.

In that way, Carl Menger exemplifies how an individual could impact culture. Founder of the Austrian School of Economics, a Professor Extraordinaire in Vienna, he tutored Francis Joseph's son, Prince Rudolf, in the economics of value, called marginalism, for six months in 1876 when Rudolph was 18, then accompanied him on a two-year tour of foreign capitals, while Rudolf was being groomed for his Hapsburg duties. Menger bolstered his own narcissism and right to influence culture by secretly and anonymously co-authoring with Rudolph a liberal, economic policy article for *Wiener Zeitung*, a government-sponsored journal (Streisler, 1990).

Living in public makes the personal life of noble aristocrats artificial and constricting. The RB Hapsburg line began to degenerate when heirs wanted to lead their own life. Rudolf's mother, the beautiful Empress Elizabeth, protected him from the stern tutoring Franz Joseph directed. Emperor of Austria from 1848 and King of Hungary from 1867 to 1916 when his assassination by Black Hand triggered World War I, Franz Joseph's reign almost traversed the time between Goethe's world and Freud's life time. Elizabeth had done her best to escape the suffocating noble upbringing which inhibited her personal freedom growing up in Bavaria. Franz Joseph married her rather than agreeing to an arranged marriage with her older sister. Allowing a fantasy of being an individual to supplant the conservative principles of an aristocratic line shows revolutionary spirit. The rift between nobility and Austro-Hungarian modernized life continued its ongoing conflict in Freud's era.

When he was 30, Rudolf deserted his own arranged marriage to a wife he did not love, for an affair with a 17-year-old. The scandal led to arguments with his father, until Rudolph took matters into his own hands by shooting his lover and himself in 1889. The noble life took away so much individuality that suicide was a way out. Searching for remnants of her life before she had been married at 16, Rudolph's mother, Franz Joseph's wife, felt she had little to live for and spent her time traveling alone until an Italian anarchist assassinated her in 1898. Modernism supported by heroic, sometimes anarchic fantasies, gradually replaced romanticized RB culture.

THE MENGER BROTHERS: SUBJECTIVE ECONOMICS

Carl Menger, born in 1840, was the third of eight children in a politically ambitious Catholic family, part of the educated liberal bourgeoisie. There is indication that his family might have converted from Judaism. Four siblings and his father died by the time he was eight. Their mother moved the family to Vienna to further the remaining boys' prospects. Carl had a two-year-older brother Max, and a one-year-younger brother Anton. We have seen that twinship experience leads to symbiotic wishes, and a culturally diffuse sense of identity. The three brothers divided their cultural roles to encompass a fantasy of shared leadership.

Carl learned to be tough-minded by competing with Max, and sensitive and creative by sharing a mind-set with Anton. After earning his law degree, Carl began to write cultural gossip columns and novels. At 25, he founded the journal *Wiener Tagblatt*, using it to publish his novel, *The Eternal Jew in Vienna* (*Der ewige Jude in Wien*). Up and coming government readers of his journal were impressed with his knowledge of how Jews used their economic understanding to establish themselves in Prussia and Vienna, and Menger accepted an offer of an editorial position in the government-sponsored journal: *The Wiener Zeitung*. Political writing for journals was often a path to building a reputation in that world. His repute showed the way for Jews like Herzl and Maximilian Harden to achieve media influence. Understanding the stock market in Vienna, Menger wrote articles on price action and value. And as a young bachelor, he used his talents and money to have many love affairs.

Menger took a LB subjective position on economic matters: The value of objects was based on their personal significance to the buyer, not necessarily on the (Marxist) labor value of production. Adding psychological and anthropological dimensions to Adam Smith's theories of capitalism, Menger is known for the concept of marginalism that calculates the subjective value of objects from the perspective of how much individuals want or need them—which may define the LB function in which money becomes a social institution (Hulsman, 2007). The libidinal value of objects was translated by Freud into the fetishistic value of cultural objects—the subjective value of objects associated with the capacity to release male strength. Let's call that economic narcissism. Perhaps Freud's economic theory of psychic energy corresponds to the energy of money in culture.

The Menger brothers had a decisive influence on Austrian economic policy through the turn of the century (Yagi, 2011). Max was chairman of the Austrian Taxation Committee from 1891 to 1896 and Anton was Finance Minister from 1893 to 1895. The three brothers divided their cultural roles according to their birth position: oldest brother as aristocrat writing about German nationalism; middle brother as the creative theorist; and the youngest, striving to be heard as an equal, writing about the democratic socialism of the common man. They all shared identification with one another that each parlayed into a unique brand of self-loving narcissism.

Many cultures are ambivalent about Jews in relation to money. Max was treated as a Jew by critics who said that his mother's lineage had been Jewish before converting to Catholicism. By 1902 Carl was living with Hermine Andermann, a Jewish journalist, who became pregnant with their son Karl (later a famous mathematician): Carl resigned his government-sponsored position, since it was against the law for Catholics to marry Jews.

As liberalism began to fail, top-down rationality eroded into a cultural regression marked by the unbinding of aggression in national groups that had been suppressed by European politics. Targeting Jews became a way to attack the political influence of liberals like the Mengers, who were accused of being Jew-lovers.

MODERNISM AND FREUD'S ECONOMICS OF PSYCHIC ENERGY

Lawrence Birken, a historian of the second half of the twentieth century studied gender, sexuality, Judaism, and economics in Freud's Vienna. Exploring Freud's economic approach to psychic energy in the context of Freud's other analytic approaches—structural, dynamic, developmental, and adaptive—he concluded that Freud attributed energy for mental work as much to cultural economic factors as to biological factors. The dynamics of sexual desire had roots in cultural experience (Birken, 1999). As I see it, the LB manages individual sexual energy, while culture manages the RB binding of aggression. Freud realized that the individual and culture are always at odds and the conflict between them creates a dynamic interaction between libidinal and aggressive neural energy.

Modernism as an intellectual movement tended to deconstruct identity in terms of money, sexuality, ethnicity, political affiliation, class, and status. The nineteenth-century breakdown in collective religious beliefs left extended kinship in the social community behind, while leaders who promoted narcissistic capitalism rationalized their power as if it were good for all. But Freud resisted modernism in the sense that he looked to archaic cultural development as one source of innate childhood development and there was romanticism, as there was in Goethe, in his feelings of cultural legacy. He believed that from the archaic world on, when grandiose leaders employed too much power to sustain their greed, cultural regression ensued.

VIENNESE MODERNISM DETACHING FROM PAST CULTURE

The cultural historian Carl Schorske provides a perspective on how turn-of-the-century Viennese modernism, by detaching from its past, formulated a present that lost its cultural emotional truth (Schorske, 1980). McGilchrist echoes those themes. Religion, and for that matter, all the other RB humanitarian disciplines looked back to the past as holistic time that had gone on forever, and which held out hope for eternal life. But the culture of modernism relied on LB logic as a sequence of goals. Narcissism is concerned with the present leading to the future. From my neuropsychoanalytic perspective, Schorske's theory relies on correspondence between LB sequential, declarative memory, and RB emotional memory. As McGilchrist says, words are merely emissaries of cultural identity, approximations of what feels emotionally real and true.

Freud, and others of his time, searched for the sexual sources of individual and cultural identity. Throughout much of the twentieth century, orthodox Freudian analysts believed that male homosexual patients used their analysis to determine whether they were truly homosexual, bisexual, or heterosexual. Often, the dynamic involved was seen as a regression from oedipal conflicts back to anal-erotic identifications with the mother, because of an earlier symbiosis with her that lasted too long and was fixating, and that in turn had led to anal eroticism conveying the wish to be impregnated by the father. Yet, Freud also concluded that many gay men had a biologically innate tendency toward feminine identity due to hormonal factors. Fading from his view was the effect of childhood sexual abuse by other men, often close relatives, triggering identification with the abuser—similar to the effect on women who identified with their attacker. Several members of Freud's circle committed suicide because they could not deal with their identity confusion. The fact that Freud used fantasy theory to exclude trauma theory shows an inclination to judge sexual identity in libidinal, LB terms rather than RB aggressive terms.

THEODORE HERZL: SEEKING HIS FEMALE HALF IN THE PROMISED LAND

Let's explore the idea that the dynamic unconscious is really a cultural unconscious. Consider Schorske's proposition that cultural modernism, alienating part of the population, triggered regression leading to World War I. Then consider Freud's idea that human development followed the evolution of cultural stages. We can view Hapsburg cultures regressing to preoedipal stages of development, disconnecting from the laws, politics, and traditions that functioned over centuries in Central Europe. Here is how Schorske depicts fin-de-siécle Vienna in a cascading regression:

> … liberals unwittingly summoned from the social deeps the forces of a general disintegration. Strong enough to dissolve the old political order, liberalism could not master the social forces which that dissolution unleashed. … The new anti-liberal mass movements—Czech nationalism, Pan-Germanism, Christian Socialism, Social Democracy, and Zionism—rose from below to challenge the trusteeship of the educated middle class, to paralyze its political system, and to undermine its confidence in the rational structure of history. (1980, p. 118)

Freud was one of the liberals who opposed Zionism. Its advocate, Theodore Herzl, born in 1860, began as a liberal who wanted Jews to contribute to progressive German culture. Like Freud, he absorbed German literature as a child. Joining the Prussian military bureaucracy, he aspired to nobility: "Herzl acquired aesthetic culture as a substitute for rank" (Schorske, 1980, p.148). Herzl's romantic liberalism had roots in his infancy, in ways the reader will recognize from the lives of Freud and Goethe: he had one sibling, Pauline, just a year older. They were so close that when Pauline died from typhus at age 19, Theodore felt he lost half his identity; he needed to shine even more. He became a dandy to show self-confidence (Cohen, 1959).

Pauline had been so central to her nuclear family's world that when she died they left Budapest for Vienna, so that Theodore, with family narcissism reinforced, could pursue his ambition to make them all—including Pauline—proud in an even more Germanic milieu. But, a highly regarded member of his college fraternity, Theodore resigned in disillusion when anti-Semitism proved as rampant among his frat brothers as it was becoming throughout Central Europe: Jews were too successful. Facing headwinds against his ambition of changing his world as a lawyer, Herzl decided to establish his reputation as a playwright. He was determined to compensate for the loss of Pauline, his symbiotic extension, by bringing Jews together in geist, just as Pan-Germans were doing; but first he had to become a successful writer, to convince Germanic culture of the value of the Jews. Marrying a pretty girl from a wealthy family and having a daughter a year later, whom they named Pauline, did not make up for failing to establish himself as a writer. His wife was no Pauline; he loved his mother more. Then his best and only close friend, who failed to succeed in the writing life they shared, committed suicide. A few months later the birth of a son, whom Herzl left uncircumcised, was no consolation.

But success came his way when he used his legal training to write about social issues for the *Neue Freie Press* as its Paris correspondent. Reporting on the Dreyfus trial, he could not believe that Dreyfus was innocent of providing French intelligence to Germany—until he realized that Dreyfus was being persecuted as a Jew. In making that transition, Herzl changed his own identity. Lauded as a new Moses by the Jewish press, he believed he would lead the Jews toward their destiny and homeland. By the mid-1890s Herzl had come to see the Jews as forced by the world to become ghettoized in spirit; they had learned to manage money as their only mode of survival without a homeland. He realized that his wish to join the prevailing culture was unworkable, that Jews needed a homeland to feel safe. Jewish nationalism became a prime example of the yearning of each colonized or marginalized ethnic group to maintain its identity. While Freud held fast to his non-Jewish and Jewish identities equally, Herzl gave himself completely to his dream of achieving the Jewish homeland—finding Pauline.

In 1895, Herzl began the 1800-page diary documenting his pursuit of the homeland:

> I have been engaged … on a work … of immeasurable grandeur … it is remarkable already as a dream … if not as a memorial for mankind, then for my own pleasure … for literature … Title: The Promised Land! (Cohen, 1959)

In narcissistic addiction to his own words, his fantasy turned toward establishing a new self of completed cultural identity. So he wrote a pamphlet, *The Jewish State,* published in a special section of the *Neue Freie Presse.* But trying to fund that state, Herzl found he was no more successful than he had been in getting his plays staged. Thwarted and humiliated by 1898, he wrote *The New Ghetto,* which when staged at the Carl Theater in Vienna became a popular play about Jewish culture. The night after seeing the play in January of 1898, Freud recorded the dream of Myops and Cyclops we have looked at, which showed concern for his children's future and the memory of his own extended family leaving for England without a home. Herzl's visionary determination to save the Jews, making him a Moses figure, stirred Freud's anger. How dare Herzl try to upstage him—like the unpleasant double in the train mirror who barged into his space?

THE EULENBURG AFFAIR

> On 23 October 1907 the case of Moltke v. Harden opened in Berlin. … It exposed a widespread homosexual network in the Kaiser's immediate entourage … key evidence came from Dr Magnus Hirschfeld. … Moltke was the military commander of Berlin, Eulenburg … ambassador to Vienna … close to the Kaiser. … Both Harden and Hirschfeld held liberal views … both opposed the Kaiser's foreign policy. … Both were Jews. (Davies, 1998, p. 866)

Following many trails suggested by Norman Davies in the above quote from *Europe: A History,* I have generated the following narrative about Freud's experience of sexual political conflicts of his time. Kaiser Wilhelm II who ruled Prussia had deposed Bismarck as chancellor decades earlier. A journalist revered by modernist writers, Maximilian Harden, published an influential journal, *Die Zukunft* (*The Future*) since 1892. He broke the news of the Eulenburg affair in 1907 by exposing a party at the Kaiser's retreat involving many of the higher echelon: the Military Secretariat collapsed and died performing a dance in a tutu. The resulting scandal focused public attention on simmering animosities about homosexuality in the Kaiser's military bureaucracy. Harden, informed years before by Bismarck of the many homosexual relationships in the Kaiser's military circle, brought homosexual behavior charges against Moltke and Eulenburg.

He wrote that hundreds were involved in homosexual activity. Prince Eulenburg, the most influential member of Kaiser Wilhelm's Junker government, had long ago used his closeness with the Kaiser to help Prince von Bülow replace Bismarck as chancellor. Moltke, was later to become the Prussian military commander in World War I. As a result of Harden's earlier exposés of rampant homosexuality from 1900 to 1906, six members of the military bureaucracy committed suicide by the time of the tutu party attended by the Kaiser.

Experts, trying to define individual identity determinants scientifically, argued about what was normal, what was pathological, what was moral, and what was lawful. The sexologist, Hirschfeld (a member of Freud's Psychoanalytic Society) settled the case by testifying that as long as homosexuals did not practice sodomy there was nothing legally wrong according to the statutes of law. Yet homosexuality was considered *inversion* or *degeneracy* by many medical experts. Freud distanced himself from Hirschfeld, a pioneer of what is now considered gay rights, who lectured widely—as far as Asia—about bisexual men as a third type of legitimate sexual identity. Freud wondered if Jews were biologically predisposed to be inverts—female-minded men, male-minded women—due to Asian heredity.

THE POLITICS OF IDENTITY

Building his base with many latently homosexual men, Freud had to be careful. For him, it was a matter of working out bisexual theories in the oedipal framework of childhood development. He had this in mind when he wrote that he feared that Little Hans would be assessed by those who wanted to harm Freud, as suffering Jewish inversion. He said Jewish men were disparaged as being like women because of their circumcision.

In 1901, Theodore Herzl published an article in Harden's *The Future* opposing Walther Rathenau's belief that degenerate Jews should convert to the prevailing culture where they could do some good. The politics leading into World War I influenced Freud's and Herzl's quandaries. Although Rathenau led the German World War I effort to centrally allocate productive resources in Germany, that did not make him less a Post-war target of the political right. He felt ashamed of what he considered his Jewish tendency to inversion.

THE SCHREBER CASE IN A POLITICAL CONTEXT

Let's consider Freud's work on *The Case of Schreber* (Freud, 1911). Schreber was appointed chief judge in the Saxon Supreme Court during Bismarck's administration. Feeling judged himself, he felt the pressure of responsibility for judging other men. Freud's work on Schreber's memoir of his illness shows how Schreber projected latent homosexuality into a delusion of being turned into a woman. Freud thought Schreber's father's sadism toward the boy set him up for the delusion that he was degenerate. Parallels between Freud's theory of libidinal regression and Schreber's individual and cultural regression are the subject of a 1996 book by Eric Santner: *Daniel Paul Schreber's Secret History of Modernity*. He says Schreber's metaphors manifest:

> a general sense of decay, degeneration and enervation … registered in fin-de-siécle social and cultural criticism. Max Nordau's famous treatise on decadence in the arts and culture, *Degeneration [Entartung]* (1892) helped to establish that term as the central metaphor for the diagnosis of cultural decline up to its fateful appropriation by National Socialist Idealogues. (Santner, 1996, p. 13)

Morally masochistic men like Schreber, sensitive to criticism about their bureaucratic judgment, yet put in the position to judge the morality of others, were primed by their father's power motifs to feel feminized by sadistic superiors. Schreber regressed when he was elevated in rank to the cultural position of judging others' moral behavior. Santner shows that the call to official power and authority are: "… *symbolic investiture* whereby an individual is endowed with a new social status" (Santner, 1996, p. 13). He says "an investiture crisis can generate … extreme alienation, anomie and profound emptiness … into the most intimate core of one's being" (p. 13).

Anal tension between obedience and discharge as one pleases were conflicting dynamics in Freud's cultural world. Like the infant's passage to young childhood, many bureaucrats, raised by their fathers with investiture expectations, may have felt relief from anxiety watching the Kaiser's drag entertainment.

Freud enacted his belief that men working together help cultural progress, by founding his Vienna Psychoanalytic Society meeting in his apartment. Cutting off from past notions of religion, tribe, and ethnicity gave modern liberals the opportunity to discover their "genius" to inform world culture how to conduct its business. Modernism gave the human sciences the opportunity to change culture by understanding it based on the biology of the mind and brain. They all believed that some chemical substance produced male and female: activity versus passivity.

Collective male narcissism focused on the leader gave Freud a godlike capacity to judge who was right or wrong. The wish to be a genius like Freud gave some members so much power of judgment that suicide could be the result. Otto Weininger exemplified that situation when as a young man of 21 he joined the Society and wrote a brilliant book entitled *Sex and Character*. A converted Jew, he wrote that women lacked the capacity to be geniuses and were to be reviled. Freud rejected it and Weininger killed himself.

WHY FREUD NEEDED FERENCZI

Sandor Ferenczi was a loyal follower throughout the development and maturation of Freud's psychoanalytic movement. Let's explore his prolonged friendship with Freud, what he taught Freud, and what he added to the movement. Ferenczi was born into a Jewish family, the eighth of eleven living children who all worked in their parents' book store frequented by Jewish, liberal intellectuals. After his father died when he was 15, his mother continued a family business meant to give Jews influence in Budapest. Since politics was the family's primary purpose, Ferenczi craved love beyond identifying with both parents. Freud was perfect: a 17 years older father figure with the influence that Ferenczi's parents wished. He joined the Society right after meeting Freud in 1908. The correspondence between Freud and Ferenczi is a rich source of understanding their relationship (Brabant, Falzeder, and Giampieri, Eds., Hofer, Trans., 1994). Like an orphan, Sandor dared not veer too far from his new family's line. In their letters, Freud addresses Ferenczi as *son*. Here is an excerpt from Ferenczi's response:

> The … heartfelt sharing that you brought to me … calmed me down … I can now think and write about Jung entirely without brother envy … he has … uncontrolled ambition, which manifests itself in petty hate and envy toward you, who are so superior to him. … His unsatisfied ambition makes him *dangerous*. … There is no alternative, you have to do everything yourself all your life. (Letter 269, p. 332, January 1912, 1994)

Reviewing their correspondence, Adam Phillips reveals their relationship (Phillips, 1994). Freud takes the position of not keeping anyone's secrets hidden except his own, while Ferenczi makes himself an open book. He cannot decide which of two women to marry, his mistress who is thoughtful, intelligent, non-demanding, and 17 years his senior, or her sexually alluring daughter whose lover had recently committed suicide; so he asks Freud to agree to analyze the daughter, in order to know who is the best bet, and Freud does so, keeping Ferenczi informed about dynamics he uncovers—without the girl knowing. They know not only her secrets, but those of others, including Emma Jung's.

Phillips remarks that Freud was the dominant male in his relationship with Ferenczi, while treating women as less important than men. He describes their bond:

> In these letters it is as though Ferenczi is Freud's repressed unconscious—the prodigal son who keeps coming back for more—wittingly and unwittingly drawing Freud's attention to the implications of psychoanalytic theory that Freud preferred to forget; partly because they were inevitably connected to his own "problems." Intimacy between people, like occult phenomena, is bewildering. (Phillips, 1994, p. 4)

Freud's advice to Ferenczi was to hold back secrets, quoting Goethe that you cannot tell boys what only a man can know. Ferenczi thought that in any event we all have similar unconscious dynamics and empathize with the emotional imagery of others that carries a conversation into the unconscious. Freud often wrote that we all have a mechanism for knowing when another person is lying. But it was dangerous for anyone to tell the leader, Freud, that he lied to preserve his power. Freud had the hubris of a mature narcissist.

WHY WE LIE TO OURSELVES: IDENTIFICATION SPLITS LOVE AND HATE

Perhaps it was the onset of pernicious anemia over a period of two or more years that released Ferenczi's caution about expressing himself openly to his patients, to the psychoanalytic world, and even to Freud. Before he died of anemia in 1932, Ferenczi gave talks based on his paper, *The Confusion of Tongues Between Adults and Children: The Language of Tenderness and of Passion*, 1933. That paper revealed the dynamics of the seductive relationship between Freud as father and Sandor as son, while explaining how Freud obscured them in his leadership role. Ferenczi set out the effects of trauma on mental process and identity from a sexual and aggressive perspective. He defied Freud's ultimatum not to publish his paper.

As we have already seen, the adult abuser equates his or her own need with the child's wish for closeness—an emotional as well as sexual violation. Feeling guilty, the child replaces a wish for love with identification. Ferenczi introduced this concept of identification with the aggressor into psychoanalysis. Melanie Klein, his analytic patient so beset with that problem in her relationship with her mother, became another source for the concept. He allowed his analytic patient Clara Thompson to kiss him, and he had intercourse with another female patient. This was revealed by another of his patients, Ernest Jones, Freud's biographer. Ferenczi used his countertransference to patients as analytic data, which in relational psychoanalysis has become common practice.

Treating many traumatized and sexually abused patients, Ferenczi concluded that they could not process their experience; instead by splitting their conscious identity they blocked the trauma, making it unavailable to interpretation. We now know that during trauma or abuse the amygdala blocks hippocampal reception of experience, substituting neutral imagery to surround the episode. We also know that trauma undermines identity, either fragmenting the feeling of self, or splitting it into known and unknown zones.

SELF-ANALYSIS: THE NARCISSISTIC SOURCE OF ALL THEORIES

Using Freud's life as a case history is instructive. One reason Freud was uncertain about the causes of his neurosis is that his identity development was split by the experience of losing part of his intersubjective self with the loss of Julius. Infant experiences of loss are hard to trace in remembered experience. Analytic work can reconstruct early life loss, by aggregating the patient's imagery to tell a story approximating the lost identity, which creates a metaphor of what really happened. Freud's splitting was so early in his life that he let others be his revenants of Julius and made a new life narrative. In our development we have at least two chances for identity formation: preoedipal identity growing out of infant symbiosis, and oedipal fantasy that reintegrates preoedipal stages. Freud always chose the path of reflecting on oedipal conflict as the narrative for his public persona. In chapter 11, we will see how Freud challenged himself to understand how women use their male identifications to moderate their tendency to be altruistic caretakers. Lou Andreas Salome played a significant role in his understanding how cultural regulation determined male and female role identifications.

Chapter 11

Gender and Surrender

Lessons in Ego Identity

Chapter 11 shows how Freud's subjective experience of building his psychoanalytic movement led him to observe how cultural conflicts affected him as well as those he was close to during and after World War I. He learned more about the effect of trauma on culture and individuals. Searching out the sources of developmental response to cultural challenge, Freud learned the distinction between male and female modes of masochism and submission revealed in regression from oedipal to preoedipal conflict. Lou Andreas Salomé played an important role in his learning.

Freud wanted to learn from Salomé how bisexual identifications of males and females affected their ego identity. Until Salomé joined his Vienna Psychoanalytic Society in 1911 when she was 50, Freud had been content with male muses. Salomé's *Freud Journal* continued her lifelong saga of creating new ego identifications with father and brother figures. She bonded sexually or relationally for pleasure, and then took the best of what each man had to offer into her own ego, which made them feel perfectly understood. In other words, her transference to men of vision seemed to benefit the ego identity of both Salomé and her partners. Her journal reflections (*The Freud Journal of Lou Andreas-Salomé*) enriched her ego identity:

> The written word gives substance to the evanescent stuff of fantasy, and at the same time is a link to the world of other persons. … For all her privileged position in the household and in society she was an isolated, lonely child, much given to daydreaming and not always able to discriminate her fantasies from her external life. … Lou Andreas Salomé, many years later, told the young Anna Freud that the only "sin" was to be untrue to one's own nature. (Leavy, 1964, pp. 4–5)

Chapter 11 dramatizes Freud's response to mixed male and female social situations. Each person has a developmental repertoire of male and female personality traits they enact differently in each cultural milieu. Freud was gallant and protective toward women, a strong male leader who used oedipal-derived wishes to show other men who was boss. After all, he was his beautiful, 21-year-old young mother's hope to better his much older father, whose leadership had been exhausted. I think Freud fell in love with Lou when she was 50 and he was 55. By this time he saw Martha as in charge of household order and cleanliness, the boss of his home; while Lou was a woman of the world who would help him bring Anna, his daughter, into his professional world—sometimes like a son.

Chapter 11 contrasts the ego identity of males and females by comparing biological and cultural sources of identity tendencies and roles. One important biological source of ego identity arises in response to the difference between stable and unstable cultural times. When collective life seems ordinary and practical, women tend to be more altruistic and pro-social than men. In the memoire of Shulem Deen presented in the Introduction, his wife chose to stay within her tribe when Shulem tried to bring his family into the larger world. Shulem's protective role as father, a source of his cultural narcissism and individuated love for the children was transformed into a legal obligation by the collective male strength of the tribe.

Inevitable conflict between tribal/religious law and the hierarchical, global economy unbinds so much aggression that revolutions and wars break out; everyone has to deal with syndromatic cultures. Does a person abide by states' rights, religious rights, subculture rights, or global rights? These questions reach deep into a person's identity, instincts for survival, and quality of life. During periods of dangerous cultural conflict everyone is prey to phantasies of recourse. In contemporary life, under influential narcissists' control of money resources, moral feelings unbind aggression in phantasies of restitution and retribution. Many feel this condition without being able to articulate it in a meaningful way. Reflection on experience is only possible if one is not detached by trauma or being a member of a cultural syndrome. Obama was a rational president who understood and communicated our global situation, but non-reflective narcissists willing to do anything to get their way as selfish leaders held sway. Chapter 11 explores the thesis that sexual identity determines social responses to challenges of well-being and survival.

HEMISPHERIC LATERALITY REGULATES HUMAN IDENTITY

Male and female lateralization tendencies in identity regulation show higher male left brain activation toward pleasure and higher female right brain activation toward the safety and well-being of loved ones. That corresponds to the lateralization of major brain-system functions:

"Both men and women showed strong asymmetries with a significant … group difference. Factor analysis on the asymmetric regions revealed 4 separate factors that each accounted for significant variation across subjects. The factors were associated with brain systems involved in vision, internal thought (the default network) attention, and language" (Hesheng, 2009).

The RB-dominant visual system provides quick access to holistic reality. RB attention, which includes salience detection, provides indications of unusual reality. The LB-dominant language mode uses verbal cues to carry out predetermined goals, while LB lateral prefrontal cortex registers failed satisfaction. Its language conveys data about failed intentions to satisfy needs, transferring that data to the RB for assessment of changes required for a new holistic conception of how to manage unusual problems. The DMN assesses whether practical thinking needs revision caused by ineffective behavior, relational changes, body changes, or cultural change. It triggers trial wishes and fantasies of identity change to determine if emotional expectations need revision.

Perhaps a greater LB tendency of men is based on male testosterone mobilizing aggression for their pleasure and dominance as leaders; while women have a RB tendency to respond to the needs of others to bind aggression. Some neuroscientists test moral dilemmas and narcissistic tendencies by assessing how innate brain networks respond to cultural conflict. *The Prisoner's Dilemma Game* is a standardized test distinguishing pro-social from pro-self individuals in their use of the amygdala and the lateral prefrontal cortex during economic choices with an unknown partner (Farmin et al., 2016). The protocol poses taking a known amount of money for oneself or a larger amount shared with a partner, which provides less than the pro-self option. It explores how brain tendencies process money sharing. While choosing, subjects' MRIs reveal network activations.

Pro-self individuals, defecting rather than cooperating, activated their left dorsolateral prefrontal areas more than pro-social subjects. Distinguishing narcissistic from altruistic people, that area, which plans behavior and follows behavioral goals toward pleasure, had more gray matter volume than the pro-socials. Addiction to pleasure at others expense is a sociopathic quality of immature narcissists. In subjects with a pro-social tendency, the right ACC activates more highly during physical or affective pain experienced by others rather than pain attributed to others (Atlas, 2010). In another study, the VMPFC linked to the ACC activates highly when modest pain triggers a *hedonic flip*: relief from worse pain (Leknes et al., 2013). They monitored the fMRI of subjects in two conditions: One signal is harbinger of worse pain; a different signal is prelude to the familiar modest pain. Modest pain is then experienced as relief/pleasure. When the right ACC signals relief to the VMPFC, the VMPFC provides context through its link to the PAG and reward systems. Relief sets readiness for renewed pleasure. Overall these studies show how male LB networks tend to be more selfish about pleasure, while female RB networks mitigate others' potential pain.

Responding to pictures of inflicting pain, sadistic male patients under treatment for sexual crimes against women showed high left amygdala activation, accompanying strong wishes to inflict pain and domination (Harenski, 2012). Their left insular cortex also showed higher activation corresponding to its trigger of sexual and aggressive desires. Another finding was high activation of the sadist's right temporal parietal junction (TPJ), known to activate during imagining another person's mental responses—mind reading. Like Ferenczi's rooster boy Arpaz, this appears to relate to the sadist's wish to know he is inflicting pain while imagining sexual domination, which Freud understood as a perverse wish to prove by sexual dominance that he is not castrated.

It appears empathy is constructed from RB network sources: the right ACC shows the same activation when one's own or another's body is a source of suffering; and the right TPJ activates when we imagine another person's mental state. Studies of the salience network show that "*the right tempero-parietal junction (TPJ)* responds robustly and consistently to sensory inputs that are novel and salient, regardless of sensory modality" (Kucyi, Hodaie, and Davis, 2012, p. 3383). The right TPJ salience network response to pain is where the RB is more strongly lateralized in activation as compared to the corresponding left area. Other nodes highly activated during pain salience are the right ACC and right insular cortex. Pain salience corresponds to RB sympathetic system vigilance (Kucyi, Hodaie, and Davis, 2012). Right TPJ salience encompasses pain as part of the affective feeling about the body as a whole entity. On the other hand, the left TPJ is a node in the attention system linked to the left dorsolateral cortex that uses inner speech to sequence behavior in the pursuit of goals: "we found that the right TPJ is negatively connected with regions in the executive control network, an opposite pattern compared with the left TPJ. This type of organization may allow the right TPJ to filter out influences of top-down information, thereby enhancing salience detection" (Kucyi, Hodaie, and Davis, 2012, p. 3383).

HOW LOU ANDREAS SALOMÉ FOUND HER IDENTITY

I want to tell the reader the story of Freud, Salomé, and Tausk as a psychoanalytically formulated narrative derived from a lot of reading and thinking. Salomé's doting father, a retired general in the Imperial Russian Army, linked his family's life in St. Petersburg to their North German Huguenot forbears.

Louise's sophisticated mother carried the adventurous Huguenot spirit. After her father died when she was 17, Louise studied religion, philosophy, and languages with a Dutch Lutheran pastor who fell in love with her, wanting to leave his wife and two children to marry her. Baptizing her as a Lutheran, he renamed her Lou. She parlayed her identification with him into erotic journal entries that dramatized her female freedom. Her mother took her to Italy and its salons, to get her away from the pastor, but left when 21-year-old Lou entered a ménage-a-trois to live with Paul Rée, a young philosopher and his friend Nietzsche. There is a famous photo of Lou holding a whip over both men in harness.

Friedrich Wilhelm Nietzsche was 38 when he met Lou. He had lost his father, a Lutheran minister, to a brain tumor when he was five years old; a year later he lost his two-year-younger brother. Just before meeting Lou, he had broken off his friendship with the composer Richard Wagner, his replacement father figure. Reflecting on his father imago led him to his philosophy that god was dead, which Lou well understood. After Lou twice refused to marry him, he wrote much of *Thus Spoke Zarathustra* within weeks, compensating for her refusal with his narcissistic philosophy—bereft of god and perfect for twentieth-century modernism. Lou turned talks with Nietzsche, Rilke, and Freud into discussions that left each feeling understood; naturally they entered a narcissistic nirvana in her female mirror.

Lou was the perfect muse for Freud. Studying psychoanalysis before contacting him, she was ready to discuss her life with him on his terms to see what each could learn. In intimate talks, using her life as their research, she helped them both clarify their perspective on the identifications each sex uses to organize its bisexual identity. Her awareness of anal eroticism as a preoedipal source of libidinal excitement and her metaphors of intercourse rejoining the broken bond between male and female identity, endeared her to Freud. Happy to bond with the men in her life, sexually or not, she let them know her body was hers, to do with as she pleased. She studied her own affective response to the quality of each man's libidinal force. Bringing that female perspective to Freud's analytic theory eased his way toward passing leadership to his daughter Anna. When Anna and her companion, Dorothy Burlingham, brought developmental child analysis into psychoanalysis, Lou was their mentor.

LOU JOINS FREUD'S DISCIPLES

When Lou joined the Society she found a band of brothers competing with Freud to supplant his leadership with their own ideas. When Freud banished Jung and Adler, Lou understood that their defections showed male oedipal competition. But Victor Tausk, 32 years old, was more original, yet more vulnerable. Lou had finished a relationship with René Rilke, the great poet whom she renamed Rainier Rilke, 14 years her junior, a decade earlier. Now she felt rapport with another younger brother figure. Tausk had come to Vienna from Slovenia and loved the South Slav heroic ballads that Rilke loved. For Lou, her relationship with Freud and Tausk was another rationalized ménage-a-trois. For Freud, Victor was a Julius fantasy transference figure.

Freud was not a leader who wanted to share a woman's love. After his last lecture of 1912: *Therapy of the Neuroses—Transference—Intellect and Affect,* he gave Lou a Christmas gift: membership to his Society—just as Lou's relationship with Tausk was heating up. In her journal entry she tells about agreeing with Tausk that adding female perspective to the analytic situation of analyst and analysand helps the therapeutic effect of psychoanalysis. Tausk said that since the sexual attitudes of the analyst affect the experience of his analysand, it was insufficient to have only a male perspective determine the outcome of analysis. She agreed, adding that since women more easily feel natural union, their sexual climax joins a man's wish for reunion with his female half. But, the relationship between Victor and Lou was neither Platonic nor secret:

> Tausk says … analytic treatment alienates the persons involved in it, while most people stress the risk of sexual countertransference. … It is not simply a question of inspecting psychic life from the outside … but … in vivo; the analyst participates actively, not just by understanding … it is the meaning of life in totality that is at stake. (Salomé, 1964, pp. 72–73)

By the totality of life Lou includes female bonding. Tausk agreed that the analyst's perspective is always implicit in his manner of listening and in every remark he makes to the analysand. Lou and Victor each enacted their own male and female transferences to Freud within their relationship, rationalizing it as a study of psychoanalysis. They realized a flaw in Freud's psychoanalysis: He breeched the necessary dyad secrecy in the analytic relationship, playing favorites over who was entitled to know the secrets kept from the rest of the members. Like a tribal leader, Freud placed himself above the rules that other members had to follow. Freud taught that resistance in the analysand defends secret sexual or aggressive feelings. He felt Tausk infringing on his entitlement. Tausk sensed that Freud needed to be the only master mind.

LOU AND VICTOR IN LOVE: WITH FREUD

When Lou met Victor in 1912, he was avoiding the responsibility of caring for two young sons and a wife, living in her father's house while he tried new careers, as a lawyer and writer, then as a psychiatrist training at Wagner-Jauregg's clinic, on the path to becoming an analyst. The oldest of nine children, he helped his mother care for his siblings while his father pursued a literary career, editing his own journal to politically impact culture—not unlike Freud. For the sake of the father's ambition the family gave up their Jewish religion, but Victor still felt Jewish. Helping his siblings, he felt like a boy in a man's role. Perhaps he could overcome that feeling by becoming a writer like his father. Freud had little empathy for him, holding that boys who imitate their father's role sacrifice themselves to their father with ambivalent over-obedience.

Tausk could not accept responsibility for helping a woman's long-term well-being, nor taking care of his younger sibs nor his two sons. He would end his relationships with women who showed dependence on him. Lou understood that Tausk lived precariously on the margin between sexual narcissism, and a failing sense of ego identity that was mitigated by his poetry and his intellectual gifts. She wanted Freud to be the father he needed, hoping to bring them together in her father/brother ménage-a-trois. She would help Freud by fitting his theory of male development into her own female stages. In a journal entry for February 8, 1913, she reports telling him she found her identity threatened in a primal scene (like the Wolf Man case):

> Awakening one night, I discovered by the feeble light that shone from the street that my father's bed, the one next to mine was empty …: he is gone … *dead* … I began to cry. Then I heard a moan from the other bed my mother … must be dying, and I let out a terrible shriek … all I could tell them amid terrifying howls was that I had become an orphan. (Salomé, 1964, p. 93)

Lou told Freud how sex reinforced her ego, and how her intercourse with Victor would improve his male ego. She was playing with fire by entering Tausk's oedipal relationship with Freud:

> Sunday afternoon, February 9, I talked about these things further with Freud. My childhood idea of woman's internal organs: like the inside of the mountains filled with precious stones. … With regard to my brother, concerning whom Freud questioned me during our evening *á deux* … when I left Russia … I looked on all the world as if it were populated by brothers alone. … It was wonderful to arrive … in the evening with flowers from Freud, to be greeted at the door by Tausk's two boys and presented with a vase full of water for the roses. (Salomé, pp. 93–94)

Being desired by men was for Lou, what a fetish may be for some men: whole body eroticism. Because Lou's erotic love and her empathy were reliving transference to her father and brothers, joining Tausk after a session of informal analysis with Freud may well have been the occasion for Lou and Victor to become lovers. The vase full of water for the roses holds erotic imagery for the two lovers. Freud had anointed Salomé *analyst*, just as the Dutch pastor gave her the identity *Lou* at her Lutheran christening. After February 9, 2013 she spent many hours at Freud's house speaking of herself and Tausk: "Thursday I was again at Freud's for supper. … Earlier in the living room he turned the conversation to Tausk and we talked a lot about him; the same later in his study and it was nearly half past one when he took me home" (Salomé, p. 97).

With his oedipal ire aroused, Freud presented *Discussions on Masturbation—Female and Male* on February 13, 1913: The lecture addressed how masturbation fantasies are mutually enacted in relationships. Salomé noted that Ferenczi was comfortable with Freud's remarks, but Tausk seemed paralyzed. Then she described what she said to Tausk later about the presentation:

> … the harm of masturbation consists in the effort to obtain the fantasy needed to take the place of a partner. … most taxing in women with a masculine tie to the clitoris … even when the fantasy is feminine in structure, I consider masturbation in women to be more injurious. … for women the sexual act is … unified with her physical and psychic being. … Only woman … knows just what it means to "cuddle" from merely resting beside each other … the night of love ministers to the day and provides redoubled energy to every task that is undertaken. (Salomé, pp. 98–99)

The following entry about a Wednesday night discussion of James Putnam's paper on American psychoanalysis brought Freud's ire to a head. It was clear enough to the disciples who knew Freud well that he was opposed to American financial and religious exceptionalism. Yet Tausk said, as if his own idea that it was an American tendency to give credence to religious speculation. After the meeting, Freud angrily took credit away from Tausk, telling Lou that Tausk was less talented than his other disciples. He also told her that when she had not been present on a previous Wednesday evening Society meeting, he found himself addressing her empty chair. No ménage for him.

World War I intervened, which we'll see from Tausk's inner eye, and then Freud's.

PRELUDE TO VICTOR TAUSK'S SUICIDE

> If all the death wishes felt by children, even if only conscious ones felt in moments of anger and hatred and released in acts of aggression of various kinds, were punished as attempted murders, the state would not have enough soldiers left to be able to set out killing its enemies. (Tausk, 1990 [1933], p. 147)

As a lawyer and psychiatrist, Tausk's responsibility in World War I was to represent Slavic and Russian deserters in court. He understood ambivalence as well as Freud did, but the childhood of the men he represented was on trial. He felt compassion for them because he knew what it felt like when your father or those in charge of culture do not care if you live or die. World War I saw casualties on a vast scale; desertion was the response of many. Tausk was always on the cusp of deserting his father—and Freud—for failing to return his love. He revealed in the above quote that children like him harboring death wishes toward their father should be saved, not punished, or sacrificed.

Tausk read Freud's work on Schreber, the judge in Bavaria, who was unable to wield his judicial power over men who may have committed crimes against the state. Perhaps he wondered if he was like Schreber. Did he also have pathological feelings about his body and mind that proved every bit of Freud's theory of psychosis? Did he too feel he would be better off castrated? Did Freud's theory that male masochism comes about in a regression to the anal period because of a wish to take the mother's place in love-making apply to him? He felt torn apart by authority, yet distrustful of his own judgment. If he did not defend the deserters, they might die.

During the war, Tausk divorced and had a series of affairs, fearing to be responsible for any woman. Preparing to return to being an analyst after his father died and the War ended, he became engaged to a young Jewish woman, an ex-patient, hoping he would be analyzed by Freud, and so become able to make a commitment to stay with her, while he rose in Freud's movement. But Freud sent Tausk to Helene Deutsch, whom he was analyzing. Helene was a doctor from Wagner-Jauregg's clinic, a younger colleague of Tausk. Freud insisted that he supervise her Tausk analysis, but she and Tausk felt that was an impossible analytic situation.

Feeling humiliated as he had been with Lou and Freud, Tausk set to work on an article that could make or break him. His famous *Origin of the Influencing Machine in Schizophrenia* (Tausk, 1919) shows his deep knowledge of how masochistic men identify with their mothers, suffer their father's reproofs, and wish to be spanked. I see the article as a response to Freud's rejection: wishing to be castrated and to feel the father's embrace as a punishment. I think Tausk felt that Freud was the real man and he, Victor, was the man without a complete body, the castrated man who could exist only through the father's love for him. He would sacrifice mind and body to Freud. He knew what it was like to feel paranoid and drained of his libidinal energy as it poured into Freud with the craziness of love. Tausk's male masochism was complete.

His article was a brilliant exposition of Freud's theory of paranoia: unable to face male masochistic wishes to be in a mother's place, the paranoid mind recoils into a virtual reality. The "influencing machine" is the modernist machine that kills, the cultural lie that treats men like children with no mind of their own. Paranoid delusions show exquisite sensitivity to cultural machines of power directed against the weak. Perhaps he felt his reproaches to Freud, like his reproaches to his father, were the reason he could not be loved. Tausk was telling Freud "analyze me if you love me." Freud was the cultural "influencing machine." In his introduction to the reprint of Tausk's classic article, William H. Sledge, a psychiatry professor at Yale, said that Tausk established the modern psychoanalytic concept of schizophrenia, for example, losing boundaries between self and other, the inside and outside, and the past and present (Sledge, 1992).

Imbedded in the footnotes of his masterpiece, Tausk signed off to suicide: Freud and Lou, Helen Deutsch, his fiancée, and his two sons would be better off without him. Freud had his family of disciples, just as Victor had eight younger children to mind as his father pursued his own purpose. In suicide he tells Freud: "I give myself to you, the cultural machine in control":

> The projection of the libido position of the psychic ego produces the symptoms of simple paranoia, the mechanism of which was discovered by Freud. In what follows we shall omit from consideration the fact that ego-libido is necessarily homosexual in its strivings. … Here I am merely calling attention to the mechanism of partial projection of ambivalent tendencies. A special contribution to the subject of the projection mechanism, which also made me aware of this principle, was made by Dr. Helene Deutsch in her discussion of this paper at the Vienna Psychoanalytic Society … Freud proposed the formulation that it is ambivalence that makes the projection mechanism possible. … The present paper shows how, albeit unconsciously, I have been demonstrating Freud's formulation. (Tausk, 1919 [1933], p. 197)

Tausk vents the reasons for his suicide, giving full ironic credit for everything he knew of psychoanalysis to Freud and even to Helene Deutsch as Freud's emissary.

VICTOR TAUSK'S EFFECT ON FREUD

Tausk suffered the fate of a Julius. He gave Freud the uncanny feeling that like a double Tausk would gain priority in new psychoanalytic ideas. Letting straw men represent his irrational wishes meant Freud never let them win the debate. Restless disciples had to go off to form their own movements. After the war, Freud felt that unexpected, life-threatening traumatic events gave everyone a jolt of the uncanny—what was overcome in the past was still happening. In *The Uncanny,* Freud wrote about the *double*, as if self were replicated by another person who had an omnipotent capacity to interfere with one's life—a person such as Victor Tausk. He resuscitated an earlier paper on the uncanny and rewrote it in the aftermath of Tausk's suicide (Freud, 1919).

The uncanny other is more than a literary device; it is a symptom of ego identity in trauma, splitting the familiar gratifying feeling of self from the fear-inducing stranger: the aggressive version of a projected self. The war brought soldiers into conflict between their commanders' orders to kill—or be killed—and disobedience. That state of mind resulted in vulnerability to trauma. In Freud's preoedipal split, Julius's death could happen all over again with a revenant—if infant Sigismund ambivalently wished it. Freud wanted to dismiss uncanny feelings as a throwback to archaic cultural times when one believed that the dead could still live. From a neuroscience perspective, the familiar quality is the LB's way of identifying what is necessary for appetitive satisfaction, for example, libidinal, fixating quality. The unfamiliar are the unknown RB entities that may relieve stress—or be a source of danger. But everyone has both a familiar sense of self and an objective state that can feel strange and even threatening— like an influencing machine. In that sense we are our own double, where death instinct comes into play.

Tausk made doubly sure of killing himself: shooting himself in the head after wrapping a cord around his neck to strangle him as he fell. Fifty years later, his son gave Paul Roazen, writing *Brother Animal: The Story of Freud and Tausk* his father's suicide note to Freud:

> … please render assistance to my beloved fiancée. … Your work is genuine and great … I was one of those who witnessed the triumph of one of the greatest ideas of mankind … suicide is the healthiest … deed of my unsuccessful life. I greet the Psychoanalytic Association, I wish it well … I hope you will have a long life, in health, strong and capable of working. … Yours, Tausk. … Please, also look after my sons from time to time. (Roazen, 1975, pp. 127–128)

There is no overt paranoia, just the irony of failed identity given over to a seeming death instinct; a death instinct quickened into suicide by rejection and humiliation from father Freud. The non-paranoid motif of altruistic surrender to another's life became the byword of Anna Freud, when she later took over the reins of her father's movement.

Roazen concludes that preserving psychoanalysis in order to save culture was more important to Freud than saving any individual, including Tausk. Freud was confounded by the role of aggression in worldwide war. Where did that leave his instinct theory? Had he been mistaken to emphasize innate libido instead of innate masochism? The failure of cultural ideals turned his theory of aggression toward a death instinct. Defending himself against empathizing with Tausk's masochistic surrender was his way of resisting his own emotional pain.

Freud's obituary of Tausk turns Tausk into a casualty of culture, a symbol of death instinct. This longest obituary Freud ever wrote presents a well-considered psychoanalytic knowledge of Tausk. It picks up on Tausk's own honesty, and screens the part of Freud's own dynamics that often remained invisible to him. It shows Freud's awareness of how culture-induced trauma degrades an individual's moral beliefs. Freud exonerates himself from abuses committed by Wagner-Jauregg on deserters (as we'll soon see)—or by himself on Tausk:

> Among the sacrifices … claimed by the war from the ranks of psycho-analysis, we must count Dr. Victor Tausk. This rarely-gifted man, a Vienna specialist in nervous diseases, took his own life before peace was signed. … during the war he threw himself … with complete disregard of the consequences into exposing the numerous abuses which so many doctors unfortunately tolerated in silence or for which they even shared the responsibility. … the stresses of many years service in the field could not fail to exercise a severely damaging psychological effect on so intensely conscientious a man. … Dr. Tausk, who has left two grown-up sons to whom he was a devoted father, was on the brink of contracting a new marriage. He was no longer able to cope with the many demands on him in his ailing state, by harsh reality. … His passionate temperament found expression in sharp and sometimes too sharp criticisms, which however were combined with a brilliant gift for exposition. (Freud, 1919, pp. 273–275)

Even if he could identify with Tausk as a victim, Freud had to stay a Gilgamesh, not an Enkidu like Jung. The obituary projects guilt for wishing to be rid of Tausk as a competitor by portraying him as too masochistic—and ornery. Freud's preoedipal identification with the aggressor led to guilt that could not be assuaged unless he saved culture from its path to destruction.

FREUD'S MORAL AMBIVALENCE AFTER WORLD WAR I

Freud believed in a death instinct from World War I evidence. Vienna had become an impoverished city, lacking solid government, no longer part of a great historic civilization maintaining sensitivity to the life of culture. When Lou's money disappeared in the Russian Revolution, Freud gave her money to live on until her analytic practice could support her and he welcomed socialist and communist disciples, many of whom were not physicians into his movement. To make matters worse, American psychoanalysis wanted analysts to be physicians. He felt America's financial imperialism had defeated German liberalism. Yet, he had to seek rich American patients to renew his psychoanalytic movement.

Freud had moral judgments to make about responsibility for the war carnage. He was appointed to head a commission judging whether Central Power psychiatrists committed war crimes by sending traumatized soldiers back to the front. Wagner-Jauregg, who taught many of Freud's disciples about organic causes of psychopathology, had been in charge of assessing German soldiers who did not want to return to battle after injury. He used electro-shock on traumatized soldiers who feared return to World War I frontlines. Freud reported:

> I am personally convinced that it was never intensified to a cruel pitch by the initiative of Professor Wagner-Jauregg, I cannot vouch for other physicians whom I did not know. The psychological education of medical men is in general decidedly deficient and more than one of them may have forgotten that the patient whom he was seeking to treat as a malingerer was after-all, not one. (Freud, 1919, pp. 213–214)

Electric shock, at high levels, had been used punitively by some physicians. Freud's ambivalent irony, his double message about Jauregg, shows he could not afford to antagonize him; many of Freud's disciples learned about psychopathology in Wagner-Jauregg's clinic. But there was a deeper reason for siding with him: he was an archaic imago of the doctor in charge of who lives and who dies, a revenant of Freud's own identification with the aggressor: the one-eyed-doctor, who held the Christian power of oedipal leader. Freud's ego identity was split between siding with the aggressor in charge—or with the helpless victim. If he antagonized Wagner-Jauregg, he would undermine the cultural status of psychoanalysis. In the same way he failed to confront those who had paid him for treating their relatives, when the guilt was theirs.

In 1927, the clash of the ambivalent Titans, Wagner-Jauregg and Freud, came to a head when Freud referred a physician to Theodore Reik, a non-M.D. disciple. Having wanted treatment by Freud, the patient sued Freud for providing him with an inadequate, "non-medical psychoanalyst". Jauregg, who had just won a Nobel Prize for curing some syphilitics with malaria, testified ambivalently for Freud, saying he had nominated Freud for the Nobel Prize—but he insisted that psychoanalysis had to be a medical discipline.

AGGRESSION AND TRAUMA: "BEYOND THE PLEASURE PRINCIPLE"

Freud wrote about trauma as a condition existing *Beyond the Pleasure Principle* (Freud, 1921). He said *anxiety* is a warning to inhibit untimely pleasure wishes; *fear* is reaction to a known threat to survival; but *fright*—a surprise fear of the unexpected—led to trauma, a repetitious condition of repeating the traumatic experience in dreams—wishing for mastery. In the Editor's note about Freud's preliminary presentation of this paper, Freud referred to Varendonck's idea that the preconscious mode of thinking (called the default domain now) deals with fright: "production of unconscious phantasies on an intensive scale in a half-sleeping state … looking ahead at the possibilities of the next day, preparing attempts at solutions and adaptations … created latent dream thoughts" (Freud, 1921, p. 5).

Freud said that daydreams after a fright induced dreams of the experience over and over. He realized that fright prepared the brain to deal with such incidents in future days:

> I think it is advisable, when establishing a distinction between the different modes of thought-activity … to designate the daydream, as well as the chain of thought studied by Varendonck, as freely wandering or phantastic thinking, in opposition to intentionally directed reflection … even phantastic thinking is not invariably in want of aim and end-representations." (Freud, 1922, p. 272)

Freud's *Beyond the Pleasure Principle* trends back to his *Project for a Scientific Psychology*. Writing as if he still believes what Fliess says about how male and female identity periods determine life span, Freud revived many old theories. Conceiving death instinct as innate masochism corresponded to Freud's own repetition-compulsion. Seeing his daughter Sophie as the return of his childhood mother, when she and her two boys Ernst and Heinele came to stay with him during the War, repeated Freud's young childhood anxiety about losing his extended family. Freud watched with fascination as 18-month-old Heinele threw an object attached to a string onto the floor and pulled it back over and over—with pleasure. Did the repetition manifest an instinct of renewing love or destroying the object? Indeed, his own lifelong compensation, and wish for mastery, created new surroundings to replace the missing part of self. His analytic movement was the transitional object of his mature ego writ large, his mythical Phantasy.

FREUD MOVES TOWARD A NEUROSCIENCE OF TRAUMA

Freud was searching for instinctual sources of trauma. Unlike anxiety, which preserves well-being through repression of impulsive behavior, Freud realized that *fright-neurosis* is triggered by surprise: the danger of an inescapable force that will destroy one's ego. So he wondered why traumatic dreams recapitulate the moment of trauma repeatedly. Does that show the "masochistic trends of the ego," or is it an excuse to take revenge? In Beyond the Pleasure Principle he writes:

> A year later, the same boy [Heinele] whom I had observed at his first game used to take a toy, if he was angry with it, and throw it on the floor, exclaiming: "Go to the fwont! " He had heard at that time that his absent father was "at the front"... he had no desire to be disturbed in his sole possession of his mother. (Freud, 1921, p. 16)

But Freud's foray into oedipal explanation of the impulse to repeat the aggression of childhood seems less relevant than anal-stage repetition-compulsion as *identification with the aggressor*. The loss of infancy by the passage to young childhood feels like the floor is being removed.

Like Goethe seeing cannon shots rending light, Freud concludes that war trauma overwhelms sensory shields resulting in the failure of perception, memory, and reflection:

> An "anticathexis" on a grand scale is set up, for whose benefit all the other psychic systems are impoverished ... the remaining psychical functions are extensively paralyzed or reduced. (Freud, 1921, p. 30) and he adds: ... the direct damage to the molecular structure ... of the elements of the nervous system ... [are] effects produced on the organ of the mind by the breach in the shield against stimuli and by the problems that follow in its train ... we ... attribute importance to the elements of fright. (Freud, 1921, p. 31)

Freud used his intellectual strength to overcome unbearable losses. Many of his conclusions, such as the following, correspond to modern neuroscience findings:

> ... the difference between systems that are unprepared and systems that are well prepared through being hypercathected may be a decisive factor in determining the outcome; though where the strength of a trauma exceeds a certain limit this factor will no doubt cease to carry weight. (Freud, 1921, pp. 31–32)

He concludes that a person may be unprepared for trauma, because anxiety has not been aroused in preparation for it. In that regard, cultural truth about the effects and expectations of war lead to preparedness and appropriate anxiety. Preparing for war by uttering kill! kill! while stabbing a dummy with a bayonet does make a killer, but also opens the mind to PTSD. Moral maturation of the prefrontal cortex helps protect a person from losing the sense of self to PTSD. But when cultural morality is surprisingly dissonant, a person's superego defenses against trauma are lost. Morally vulnerable people may become sociopathic to resist traumatic threats to survival.

Caught in inescapable trauma, RB limbic cortex signals the right amygdala to trigger emergency responses, mobilizing a cascade of neural events that block hippocampal reception of new experience. Instead, the amygdala stores affect-images associated to the life disaster. When those images recur they present an affect of recurrence, like the multi-headed Cerberus dog that bars the entrance to personal hell. The familiar self recedes from reflection and falls out of view.

FREUD'S ULTIMATE COMPENSATION FOR HIS LOSSES: ANNA FREUD

After World War I and the death of Sophie, Freud's youngest daughter Anna grew from an adolescent into a young adult determined to become an analyst like Lou. Freud began to analyze 26-year-old Anna in 1921, stopping in 1923 after an operation on his soft palate cancer. Once he recovered, he analyzed her for two more years while she went through analytic training. Like many women of her time, Anna had not gone to a university. But her father defied American-led insistence on medical training for analysts, setting a precedent for child analysts to have their own institutes and training. We learn from Anna Freud's biographer, Elisabeth Young-Bruehl, that by 1921, at Freud's invitation, Lou became Anna's supervisor in psychoanalysis (2008).

Young-Bruehl's biography reconstructs Anna's place in her family's mythology. Her birth so overwhelmed Martha that she went to stay with her own mother in Germany for several months to steady herself. A Catholic nursemaid was hired in her place, who became devoted to Anna. Anna later referred to her as her *maternal caretaker,* the term she used for herself in her own role of establishing and running nurseries. Hiring a caretaker devoted to baby Anna more than to the other children; Freud recapitulated replacing his own depressed mother by bonding to his maternal caretaker, Monika. Anna's birth seemed to mark the end of her parents' sex life.

Freud's description of baby Anna's dream points to her infant inkling of ego identity:

> My youngest daughter, then 19 months old, had an attack of vomiting one morning and had consequently been kept without food all day. During the night after this day of starvation she was heard calling out excitedly in her sleep: "Anna Freud, stwawbewwies, wild stwawbewwies, omblet, pudden!" At that time she was in the habit of using her own name to express the idea of taking possession of something. The menu had included pretty well everything that must have seemed to her to make up a desirable meal. The fact that strawberries appeared in its two varieties was a demonstration against the domestic health regulations … she … no doubt observed that her nurse had attributed her indisposition

> to a surfeit of strawberries. She was thus retaliating in her dream against this unwelcome verdict. (Freud, 1900, p. 130)

This infant dream, which Freud mentions three times in his *Interpretation*, also contains the base of Freud's mythology of his own childhood. The nursemaid replaces mother, giving the infant too many strawberries and then makes amends by withholding them. Could Freud avoid musing on *his* caretaker after Julius died of intestinal problems? Anna enters into his mythology.

Family mythology to screen emotional losses exerts a strong cultural influence on the identity of every child, which then plays a part in his or her lifelong fantasy formation. Anna had a capacity to turn someone else's resentment aside by understanding the other person—she was brains before beauty in her family mythology. Entering as a burden to her parents and siblings set her up as the one who must make altruistic surrender, which became her status in the analytic community. Defending against instinctual satisfaction in favor of gaining ego strength through identifications gave her a different path from object-relations analysts like Melanie Klein, who dealt more directly with the raw aims of human instincts.

Young-Bruehl's history of Anna's family presents a female perspective on Anna's home milieu. Her mother Martha, her mother's sister Minna, and Josephine the nursemaid ruled. Young-Bruehl reports Anna's masturbation fantasies of being beaten in her five-year-old stage. She then had conscious fantasies of a boy being beaten during her latency. Since no child in Freud's family was physically beaten, the memory may condense what she witnessed in school. Corporal punishment was still part of school discipline, especially for boys. Her oedipal fantasy of being beaten by her father during clitoral masturbation had been repressed to her unconscious:

> … the creation of the beating fantasy … itself a substitute for an incestuous father-daughter love scene that "distorted by repression and regression to the anal-sadistic phase finds expression as a beating scene," the climax of which coincided with masturbatory gratification. These fantasies appeared before the girl entered school … and they continued until—between her eighth and tenth years—they were replaced by what she called "nice stories." … (Young-Bruehl, 1988, pp. 104–105)

Young-Bruehl says Anna identified as a male to avoid penis-envy anxiety and her analysis confirmed that oedipal regression to preoedipal conflicts happens in girls as well as boys. Writing about Anna's case as if it were another patient, Freud made her an exemplar for female childhood development, as Little Hans had been for male development.

ANNA FREUD CLARIFIES THE LOGIC OF PSYCHOANALYSIS

After 1923, the way to Freud's heart was to help Anna replace him. Males who did so received signet rings: membership to his inner circle; women were given brooches as proof of devotion to spreading psychoanalysis. With the help of Heinz Hartmann, who was a psychiatrist at Wagner-Jauregg's clinic and who would become Freud's emissary of ego psychology to America, Anna went on rounds at the clinic. She and Hartmann learned that both emotional trauma and organic damage undermine ego identity, and sometimes grease the skids for psychotic regression.

Anna's ego identification with her father's rational purpose and her mother's control of everyday life shaped her identity. The outcome of her analysis was an altruistic surrender of her fantasy life to the cause of psychoanalysis. After 1923, Anna and her father worked together to make his structural theory into ego psychology: id, ego, and superego. Anna Freud personified ego psychology. She sublimated oedipal fantasies and latency daydreams of romantic fulfillment into an adult capacity to help children. She founded developmental, ego psychology, based on child analysis. Writing in her journals about her own development and her own defenses like Lou reflecting in her journals, she realized that identity-organizing sexual fantasies organize ego identity. She formulated defenses used by the ego to maintain practical consciousness.

Emulating her father, she used her own case as the source of the papers she wrote to qualify for membership in the Vienna Psychoanalytic Association. Young-Bruehl's overview of her analysis with her father used data from Freud's paper, A *Child Is Being Beaten* (1919) and Anna's paper, *Beating Fantasies and Daydreams* (1922), which she presented for graduation. Pretending her own case was someone else's, she followed her father's tradition of using himself as the template for all patients through the artifice of writing about himself as another person: the imposturous phantasy as mental creativity.

Yet Freud felt stress about yielding leadership of his analytic movement. He wasn't so sure that female masochism was altruistic. He learned from Helene Deutsch, as well as Anna, that a woman's psychology is not parallel with a man's development. By the time he resumed Anna's analysis in 1923 and then wrote *The Economic Problem of Masochism* in 1924, Freud's theory of instinctual development included a primal instinct of masochism. But he did not know how to describe bisexual ego identity development as it differs in men and women. He tried by recognizing his counter transference as he analyzed Anna in *A Child Is Being Beaten* (1919):

> The affections of the little girl are fixed on her father, who has probably done all he could to win her love, and in this way has sown the seeds of an attitude of hatred and rivalry towards her mother. This attitude exists side by side with a current of affectionate dependence on her, and as the years go on it may be destined to come into consciousness more and more clearly and forcibly, or else to give an impetus to an excessive reaction of devotion to her. ... (Freud, 1919, p. 186)

Freud's limitation lay in not distinguishing female masochism from male perversion. He wondered if primal masochism is a tendency to bind aggression through intrinsic morality. He solved the dilemma with structural theory. In ego psychology, superego control of aggression is called *binding*. Since moral interaction is how Anna pursued her life work and relationships, we will next see how it shows up in Freud's *The Economic Problem of Masochism,* 1924. He wondered if Anna's *altruistic surrender* was an empathic quality of satisfying the cultural aims of others, or manifested instinctual primal masochism. Anna also had the same doubts as her father.

THE ECONOMICS OF MASOCHISM

Freud was pursuing correspondence between basic instincts and the development of mental feelings, by aligning instinctual duality with conflicting emotions. He equated pain and pleasure with aggression and libido. From early infancy the libidinal side of ambivalence is engendered in the brain's tendency to bond with familiar caretakers—while an aggressive response to the unfamiliar stranger brings anxiety with emotional pain. The ACC activates highly in six-month-old infants in response to caretaker conflict! (Graham, Fisher, and Pfeifer, 2013).

Freud envisioned evolution as proceeding from unicellular creatures through the whole organic kingdom. He assumed that physical principles of reality coevolved with social qualities: "... the Nirvana principle, belonging ... to the death instinct, has undergone a modification in living organisms through which it has become the pleasure principle ... the life instinct, the libido ... seized upon a share in the regulation of the processes of life" (Freud, 1924 pp. 161–162).

As I now see it, the two hemispheres can reach different conclusions about the meaning of experience. Perhaps that is the source of both ambiguity and ambivalence. Just as moderate pain can become pleasurable in a bilateral identity context, masochism can also be altruistic and pleasurable in a cultural context of survival of all the parties concerned. If cultural identification mediates aggression, then guilt can work toward survival of everyone. In female identification, Tausk could well have been expressing his altruism. Freud goes on about superego identification:

> The course of childhood development leads to an ever-increasing detachment from parents, and their personal significance for the super-ego recedes into the background. To the imagos they leave behind there are then linked the influence of teachers and authorities, self-chosen models and publicly recognized heroes, whose figures need no longer be interjected by an ego. (Freud, 1924, p. 168)

That speaks to how it is that internet heroes for Isis can enter the phantasies of the down-trodden and sociopaths as if they were moral figures by offering pleasure without responsibility:

> … that moral masochism is unconscious leads us … to translate … "unconscious sense of guilt" as meaning a need for punishment at the hands of a parental power. We now know that the wish … appears in fantasies … to be beaten by the father stands very close to the other wish, to have a passive (feminine) sexual relation to him. (Freud, 1924, p. 169)

Freud's love for Anna promoted her wish to altruistically surrender her instinctualized identity to help him lead culture. Her identity featured primary masochism *and* primary empathy. Chapter 12 will move us from the sympathetic altruism of Anna Freud toward the élite narcissism that influenced American psychoanalysis leading up toward World War II.

Chapter 12

Freud's Ambivalence about America

Chapter 12 looks at the changes in Freud's theories and cultural outlook in the decade after the war. Creating psychoanalysis as his life work, Freud was influenced by his cultural experiences, which led to further reflection on his own identity changes in the course of his life. I imagine that Freud's guilt over breaking off from Fliess and Jung—leaving them in his lurch—and his role in the death of Victor Tausk, who meant to satisfy Freud with his suicide, and the reexperience of poverty and social and cultural loss in World War I, all fueled Freud's Julius guilt, turning his reflection onto how culture ruins the brain. Culturally diverging theories of psychoanalysis divided analysts in their understanding of what was most basic in psychoanalytic clinical approaches. The social historian Elizabeth Ann Danto documents Freud's turn to liberal socialism after World War I:

> Freud supported free psychoanalytic clinics, stood up for the flexible fee, and defended the practice of lay analysis, all substantive deviations from a tradition of physicians privilege and their patient's dependence. His consistent loathing of the United States as "the land of the dollar barbarians" echoed his contempt for a medical attitude he believed to be more American than European, more conservative than social democratic. This … resulted partly from the grievous material and psychological deprivations the Freud family endured during the war and partly from momentous shifts in the larger political landscape of the early twentieth century. (Danto, 2005, p. 13)

In his *Lines of Advance in Psycho-Analytic Therapy* paper, Freud established a new psychoanalytic plan for the future. In a first post-war conference in 1918, he set out the theme:

> … the poor man should have as much right to assistance for his mind as he now has to life saving help offered by surgery … institutions or out-patient clinics will be started, to which analytically-trained physicians will be appointed … men who … give way to drink, women who have nearly succumbed under their burden of privation, children for whom there is no choice but between running wild or neurosis, may be made capable. … Such treatments will be free. (Freud, 1918, p. 167)

With the financial help of a rich Hungarian analyst, Anton von Freund, Freud issued vouchers for psychoanalysis, to be used interchangeably among all the analysts who participated, even students working in the 10 free European clinics he authorized.

Looking back on that decade in *The Future of an Illusion*, and looking ahead to the future Freud wrote about the regressive cultural effect of leaving the poor to suffer:

> … underprivileged classes will envy the favored ones their privileges and will do all they can to free themselves from their own surplus of privation. … Where this is not possible, a permanent measure of discontent will persist within the culture concerned and this can lead to dangerous revolts … it is understandable that the suppressed people should develop an intense hostility towards a culture whose existence they make possible by their work, but in whose wealth they have too little share. (Freud, 1927, p. 12)

FREUD'S AMBIVALENCE ABOUT AMERICAN MONEY

Compared with the confident American optimism of unfailing narcissism, Freud was frustrated with post-war European poverty. He still had to rely on wealthy American patients—for money to maintain his journals and to administer his institutes. Freud's ambivalence about America stayed negatively focused on President Wilson through his political relationship with William Bullitt. Besides, he was still permanently rankled that Ely Bernays, Martha's older brother who migrated to the United States in 1892, had withheld money left to Martha by her father, which could have shortened Sigmund's and Martha's eight-year wait to marry. Unlike Jacob, Ely was a rich grain merchant who had married Freud's sister Anna—born after Julius died. Yet Ely's son Edward, who had studied Freud's work, brought Freud the gift of his influence on America.

Visiting Ely's family on his 1909 trip to New York, Freud was drawn to Edward. As an ad-man, Edward used Freud's insights to influence consumers. For example, he promoted a women's rights demonstration: female models smoking Lucky Strikes paraded on New York's Fifth Avenue as *Torches for Freedom.* His status as "father of public relations" was such that President Wilson brought him to the World War I Peace Conference to promote the League of Nations. Visiting Freud in Vienna on that occasion, he brought him a box of Havana cigars, endearing him to his uncle forever. Freud still harbored wishes to have moved to England with his stepbrothers, to have avoided poverty. Cigars helped him feel detached from that poverty.

Freud was increasingly distrustful of America as some of his followers migrated to America where they were able to make more than a good living analyzing the rich and famous. For one thing, orthodox American psychoanalysis insisted on medical training as necessary for psychoanalytic license. Freud had little empathy for those who moved to America where the lure of opportunity added a higher narcissism to their ego identity. Freud felt they bought into American exceptionalism. Yet Dorothy Burlingham who became Anna Freud's later life partner and William Bullitt, both influential patients of Freud's in the early 1920s, were American analysands with clout and money who improved Freud's life.

THE STRUCTURAL THEORY BRINGS CULTURE TO MIND

In 1923, Freud's structural theory of id, ego, and superego brought culture as superego into the mainstream of psychoanalysis. He designated id as the biological source of mind, ego as the source of individual and social identity in the mind, and superego as regulating cultural identity. As a guiding member of the Psychoanalytic Society by 1925, Anna Freud used her own ego identity, identified with her father's rationality, to redefine psychoanalytic goals. Her method in adults and children was analyzing defenses that blocked individual maturation. Heinz Hartmann, who later moved to America, viewed superego regulation from a cultural perspective. He conceptualized *Bound aggression* and *bound libido* as manifesting the energy of bound instinctual drives. Bound libido pursues a moral ideal of sublimation for the sake of collective problem solving; bound aggression uses moral rules to help the individual feel concern for others' suffering. Superego bound aggression protects others—a sign of compassionate kinship.

DIVERGING PSYCHOANALYTIC FAMILIES

Hartmann's cultural perspective clarified the division between Anna Freud's and Melanie Klein's approach to psychoanalytic practice with adults and children. Promoting mature narcissism, Anna Freud's version of *Self-psychology* engaged followers such as Kohut, while *object relations* analysis dealing with internalized aggression appealed to Kleinians like Otto Kernberg. Their clinical approach interprets overly raw aggression whenever it comes up, while Kohut's self-psychology points out the wish for self-mirroring pleasure. Kohut treated narcissism as a normal developmental mode based on libidinal identifications; while Klein analyzed identification with the aggressor.

Analyzed by Ferenczi in Hungary during World War I, Klein became a child analyst after the War, including her son and daughter among her analysands. Then she had further analysis in Berlin with Karl Abraham. In *Freud and His Followers* Roazen presents the Anna Freud, Melanie Klein "divide" (Roazen, 1975). Invited to England by Freud's biographer Ernest Jones, Klein became popular as an analyst of analyst's children. Roazen notes that by winning Jones over, she overcame her envy of Anna Freud. After all, Freud had imagined Jones as a possible suitor for Anna. Klein led a group of *object relations* analysts that came to include Ronald Fairbairn, Donald Winnicott, and Otto Kernberg. She concluded that infants experience part objects like breasts or faces, rather than the whole person, before they assembled parts of a whole person. The breast could be a good or bad object. Klein suggested that traumatic identification with aggressors in infant life led to pathology in binding aggression and managing empathy.

Using children's play in place of the free associations of adults in her treatment, Klein suggested that oedipal conflicts begin in infancy when love and hate contend, then continue in anal-stage conflict manifesting as tantrums of envy bound by abandonment anxiety. Freud often considered conflict from a Kleinian perspective, only to shift to oedipal reorganization at the end of most of his papers. Klein followed Freud's original way of working by interpreting instinctual sources of behavior. She saw raw instinct in children's play as creative fantasy foreshadowing development; just as Freud, like his followers, saw creativity as regression in service of the ego.

From my neuroscience perspective, the RB integration of a structured, objectified, whole self is compromised in borderline individuals. Their internalization of culture is impaired at each stage of development, so binding raw instinct through identification is incomplete. That process requires both intact brain networks and good-enough parenting in early childhood. Donald Winnicott's work on the "false self" of borderline patients shows the problem of a failure to bind aggression and libido in the formation of an integrated personality (Winnicott, 1965). The prevalence of raw aggression and raw libido shows failure to develop lasting identifications and normal ambivalence in the anal stage.

Looking at Kernberg's work from that context (Kernberg, 1975), LB good objects are experienced as a symbiotic extension of the pleasure drive; while RB bad objects invade the RB's reality domain evoking painful affects due to unbearable frustration. In my view, the RB uses bound aggression to integrate a holistic sense of real objects. But borderline people incorporate the aggressive object in infant identity causing so much stress that a syndrome develops to split bad from good objects. Kernberg says ambivalence does not develop normally when the bad object has invaded the inner sense of identity. Object relations analysts focus clinically on binding aggression through establishing realistic ambivalence toward the analyst: so the patient can stop projecting aggression as blame, which splits others into good versus bad objects. Bias against multiple cultural identities unbinds aggression like terrorists in a Trojan Horse.

ANNA FREUD'S FAMILY OF THE HEART

Anna Freud followed her father's cultural tradition of mature narcissism. By 1925, she shared the reins of psychoanalysis with him. Her goal for child analysis was to teach children how their feelings about their parents and caretakers helped them grow up. That meant they had to learn how their feelings determined their behavior. Finding out how affects blocked self-expression was part of their emotional education. She studied how instincts transfer their energy to ego identification through the work of defense mechanisms. As we saw in chapter 11, she used reaction formation to turn her own sexual and aggressive fantasies into more socially acceptable ways of enhancing her ego identity. By analyzing a child's defenses against knowing what he or she wants and how to get it, Anna Freud helped children enhance their self-esteem and their mature narcissism. Perhaps her way encourages elitism.

As we see in his comments on *Wayward Youth* (Aichhorn, 1951), the work of Anna's follower, August Aichhorn, Freud encouraged Anna's approach to ego psychology:

> ... children have become the main subject of psychoanalytic research. ... Analysis has shown how the child lives on ... in the sick man ... in the dreamer and the artist ... and traced the stages through which a child grows to maturity ... psychoanalytic concern with children will benefit ... education ... to guide and assist children ... shield them from going astray. (Freud, 1925, p. 273)

Aichhorn was a Catholic educator whom some call the father of social work. After his twin died when he was 20, he used his symbiotic transference in working with patients. He related to delinquent adolescents as if he knew their subjective wishes. He went along with them far enough to get them to idealize him; then he had the upper hand in helping them choose socially acceptable ways of getting what they needed. When August and Anna were working as educators in Vienna, Anna befriended him, and with Freud's help arranged for his analytic training at the Vienna Psychoanalytic Society. Freud's affection for Aichhorn shows in his preface to Aichhorn's classic *Wayward Youth*:

> [Aichhorn] has learnt analysis by experiencing it on his own person and … being able to employ it in borderline and mixed cases … he should obviously be given the right to practice analysis, and narrow-minded motives should not be allowed to try to put obstacles in his way. (Freud, 1925, p. 275)

Aichhorn supervised Heinz Kohut. His intersubjective style shaped Kohut's Self Psychology. Kurt Eissler lauded Aichhorn's way of inviting subjective identifications in his preface to Aichhorn's *Wayward Youth*: "… he could implant a message in the unconscious of his subject … his questions were directed to … create certain preconscious associations … he had forged an exquisite therapeutic tool out of a procedure which was … routine" (Eissler, 1951).

Like Aichhorn, Kohut used impersonation to show empathy for his patients, Charles Strozier, Kohut's biographer, revealed that Kohut's famous *The Two Analyses of Mr. Z.* was about himself (Strozier, 2001). Like many famous analysts, beginning with Freud, Kohut used his own life story and his analysis of it as a published case presented as some other person. He grew up with a mother who wanted to know and control everything he did. In his teen years she hired a tutor who became his only friend and source of knowledge about the world. Strozier shows that was exactly the situation of Mr. Z. Kohut concocted an ersatz first analysis of Z, a model analysis complying with Freud's standard view of acknowledging oedipal themes. The so-called second analysis of Mr. Z corresponds to Kohut's analytic work with Aichhorn. Z realizes that his mother's overwhelming love for him was self-love, leaving no room for autonomy, so he learned that he was mistaken to have thought his father lacked qualities he could identify with.

Phyllis Greenacre, an analyst and psychobiographer of creative people, explored imposturous qualities, of pretending one's identity belongs to someone else, a subject fascinating Freud (Greenacre, 1971). She added to Freud's work on the family romance fantasy as the cultural core of a creative person's belief that he or she can impact culture. Greenacre concluded that fantasies of alternative parents, parental overestimation of a child's capacities, fantasy used to enhance self-esteem by elevating ego identity, fixation to fantasies to relieve painful body afflictions, and early childhood emotional losses and emotional trauma were all factors in developing narcissistic imposture. I was particularly impressed with her observation that the sense of imposture often comes about through early childhood experiences of inner-body alteration in response to emotional stress, such that odd body feelings were projected onto the outside world. She examined how much creative artists anthropomorphize the external world. Experiencing the outer and inner world as symbiotically fused seems like uniting with god.

THE AMERICAN GUEST AND HER FOUR CHILDREN STAYED FOR PSYCHOANALYSIS

In 1925, Dorothy Tiffany Burlingham came to Anna Freud for help in finding a new way of life for her four children and herself. Anna analyzed each child, cobbling together a school for them and a few other children of analytic-minded families; while father Freud analyzed Dorothy, who became part of their family. Young-Bruehl said Dorothy wanted to be a twin like her older sisters, so as a child she invented an imaginary twin (Young-Bruehl, 1988). She was the last child of parents wanting a boy who had four girls. Young-Bruehl thought Anna and Dorothy both had family mythologies that added a good dose of male identification to their character structures. Their relationship gave them each a needed satisfaction of mirroring friendship.

Identifying with their fathers, Anna and Dorothy expected that they too could impact culture. Their educational work for the City of Vienna in the 1920s, their analytic study of child development, and their work in London's World War II Hampstead War Nursery led to an improved awareness of what children needed in order to develop their potential. Seeing abandoned wartime infants failing to thrive when all of the caretakers took care of all the infants, they assigned just a few to each caretaker. Symbiotic bonding showed that one caretaker in the first months of life diminished effects of hospitalism. An infant needs mothering to trigger symbiotic relationship.

Erik Homberger, known as Erik Erikson and for the phrase "identity crisis," and Peter Blos staffed the school set up for Dorothy's children. Anna and Dorothy benefited from identifying with their fathers' wills to impact culture, while Erik and Peter benefited from identifying with their mothers' empathic capacity for creative fantasy exploration of what it means to be human. Erik was the son of a lover (Erik's pediatrician) his mother had while waiting for an official divorce from her husband. Unsure of his identity, Erik changed his last name to Homberger, that of his biological father. But in analysis with Anna, reborn into his own identity, he changed his last name to Erikson.

DOROTHY TIFFANY BURLINGHAM'S AMERICA

Dorothy Burlingham, an American heiress, bolstered Freud's and Anna's confidence as cultural leaders. Beyond the analytic fees she paid the Freuds in times of their need, she also contributed an American ethos. Imagine you are Freud analyzing Dorothy as she tells how her wealthy father's libidinized art fed America's narcissistic consumerism. Imagine her driving the Freud family to fashionable places when others had no car. Freud's narcissism always expanded when he analyzed exceptional Americans.

Dorothy's grandson, Michael John Burlingham, wrote an entertaining biography of her life: *Behind Glass* (Burlingham, 2002). Born in 1812, her grandfather, Charles Lewis Tiffany, amassed a fortune as the premier gem dealer for wealthy Americans intent on outdoing Europe's aristocrats. He learned the art of consumer persuasion from Phineas Barnum, who advised him to advertise that he surpassed European quality with his American jewelry. Tiffany boldly bought the gems of French aristocrats in 1848, when a second revolution sent the remnant royalists packing for good, and sold their jewels to upper-class New York shoppers at a fair price. Showman of American exceptionalism after the Transatlantic Cable was laid in 1857, Tiffany bought the leftovers, packaged four-inch pieces into jewelry, and sold them as souvenirs.

In 1909, when Dorothy was 18, Freud's New York visit began a psychoanalytic ethos in America. Dorothy, all her sisters, and many friends were intrigued with Freud's ideas. Imagine Freud's response to Dorothy remembering his visit to New York, days before his Clark University lectures, using some of his $7500 speaker's fee to buy a jade bowl and a bronze Buddha at Tiffany's. Freud also acquired a bronze porcupine that he kept bristling on his desk, a gift from James Putnam, a renowned Boston neurologist who lauded Freud's work as a humanitarian approach that reduced stress on the psyche. To avoid having his movement identified as Jewish, Freud encouraged Putnam to establish an American Psychoanalytic Society.

Dorothy's father Louis Comfort Tiffany was determined to make American stained glass lovelier than European glass, trying many techniques until he succeeded. His Tiffany Glass Company was staffed with well-paid women who manufactured his stained glass creations with mosaic precision. After his first wife died, leaving two girls and a boy, he married Dorothy's mother. Louise had twin girls, a boy who died, and Dorothy who seemed to replace the boy. After Louise died when Dorothy was 13, Louis built a dream mansion at Cold Spring Harbor on Long Island. His party on February 4, 1913 shows Dorothy's early adult American life:

> The Havermeyer, Macy, Fisk, Roosevelt and Rockefeller families were all represented. John D. Rockefeller was a Persian prince; his wife, Minerva, the educator Horace B. Mann, a Roman senator, Albert and Henry Herter were Egyptian noblemen; Theodore Steinway, a Roman soldier … mortally embarrassed Dorothy was obliged to refresh the great Queen with an enormous peacock feather fan

> … dancer Ruth St. Denis was unrolled from a rug to dance half nude … the guests dined on an eight course dinner catered by Delmonico's. Enough cases of St. Marceau Brut, 1906 were imbibed to fuel the after-dinner turkey trot … themes of sexuality and slavery gave way to … inebriated revelry … behind the historical pageantry and celebration of cultural opposites … lay a … personal allusion to Louis's newfound happiness with Patsy. … (Burlingham, 1989, p. 130)

CULTURAL CONFRONTATIONS IN FREUD'S DOMAIN

In 1914, Dorothy married the son of an important American; but Robert Burlingham was a mere shadow of his father, Charles Culp Burlingham, called *CCB*. Because CCB confronted Freud in a way that threatened Freud's new analytic family let's get to know him: Phi Beta Kappa from Harvard, a Columbia law graduate, and a power behind the liberal New York City and State politics, he came to fame representing the White Star Holland American Line that owned the sunken Titanic. The U.S. Supreme Court accepted his argument that in terms of indemnity, U.S. regulations preceded any other authority. President of the NY City Bar Association, he helped Learned Hand and Benjamin Cardozo win appointments to New York State Courts. He ran LaGuardia's 1934 campaign for Mayor, ending the reign of Tammany Hall in New York City.

Like Woodrow Wilson's father, CCB was a religious Episcopalian, praying over every decision, expecting his son to do the same. Robert and Dorothy had four children before it became clear that Robert was not his own man. When he had to make decisions, about working as a doctor or about family life he became so agitated he required hospitalization. When their first son Bob seemed to be developing Robert's emotional instability, Dorothy searched for a way out of her predicament. By 1921, after the birth of her fourth child, she left Robert, took the children, and hired nannies to care for them. In 1925 she moved to Europe with the children, putting them briefly in a school in Zurich run by psychoanalytically informed friends. Hearing that Anna Freud analyzed children in Vienna, she decided that a few months' treatment for Bob might help him.

When it became clear that Dorothy and the children were living with the Freuds, CCB pressed Robert to get the children back to New York. Together, they challenged Freud and Anna. Robert and CCB met with the Freuds, demanding to know the health of Dorothy in analysis with Freud, and of the children in analysis with Anna. First Robert interviewed Anna, but he could not get her to say that his presence was better for the children's development. The next day he asked Freud if he meant to interfere in his and Dorothy's sex life: Freud said he never came between husband and wife—Dorothy had a mind and body of her own. The next day CCB questioned whether Freud feared he and Robert might kidnap the children. "Were they not gentlemen," he asked? Freud answered with relish that Robert was a gentleman, but CCB was a lawyer. CCB and Robert returned to New York empty-handed.

THE FREUD WARS IN AMERICA

Looking back on the twentieth-century culture of Freudianism, Young-Bruehl, suggests that post-modernists debunked Freud's work to show that they were the new elite (2008). She observes that intellectual competitors for Freud's leadership often claimed that he denied the effect of abuse in his patients as a cover-up in order to emphasize the efficacy of his psychoanalysis. These accusations, Young-Bruehl concluded, were often a rationale for competitive wishes, rather than respecting his therapeutic work that implemented cultural progress.

Jeffrey Masson, who became project director for the Sigmund Freud Archives in 1980 under Kurt Eissler's direction, used his access to classified material to claim that Freud dropped his seduction theory of incest either because he wanted the abusers to pay for the analyses he conducted or because he wanted his fantasy theory to gain acceptance. Those accusations deny the complex dynamic aspects of Freud's motivations, as well as the fact that he put the health of his movement in front of any individual's well-being. Young-Bruehl wondered if this *Bad Freud* story was derived from Masson's own oedipal situation of growing up with a father who taught him that it was in their blood to seduce women. Was narcissistic compensation Masson's reward for blaming Freud? Eissler had thought when he vouched for Masson that like himself, Masson had overcome his self-centered approach; so he failed to warn Anna that Masson was not just a good scholar well-versed in psychoanalysis, but a master of innuendo denying his own motives.

Young-Bruehl cites Christopher Lasch's *The Culture of Narcissism* as describing those like Masson who sought American prominence by waging the Freud Wars:

> In a narcissistic society—a society that gives increasing prominence and encouragement of narcissistic traits—the cultural devaluation of the past reflects not only the poverty of the prevailing ideologies, which have lost their grip on reality and abandoned the attempt to master it, but the poverty of the narcissist's inner life. (Lasch, 1979, p. 13)

There is a substantial difference between grandiosity and mature narcissism. Grandiose narcissists steal their identity from past minds. Some deconstructionists hold that a published dream never did belong to the dreamer, but to cultural associations of the reader. Jacques Derrida, for instance, takes all writing as a means of imposing control over others. Deriding the writer's subjectivity is like throwing out the baby with the bathwater. But even more brazen, contemporary writers increasingly use past writers' words as their own. Freud thought the Bolsheviks used their immature narcissism to claim that the masses could not think for themselves, and therefore they had to think for them. Marx's words became a convenient rationale for those who wanted to exert cultural control with seeming compassion.

Self-psychology is a romantic pathway of psychoanalysis using an understanding of subjective development to gain mature influence. Many psychoanalysts of the late nineteenth and early twentieth centuries, including Anna Freud in her identification with the Cyrano de Bergerac role, created their own poetry, fiction, or drama (Young-Bruehl, 2008). There is a romantic, archaic quality in many people who use fantasy for the sake of creativity. Imbedded in romance is a wish for symbiotic fusion with other minds, which is the essence of narcissistic intersubjectivity. Freud wished his identification with forebears like Goethe were links to life forces and archaic culture. Denying telepathy and oceanic feelings in his writings, yet absorbed by those wishes, indicates the subjective side of Freud's splitting. Despite their respect for many of Freud's male oedipal attitudes of logical superiority to helpless symbiotes, Anna and Dorothy brought female perspective to their work, enjoying the feeling of relational life together.

NARCISSISTIC AGGRESSION IN CULTURE

Immature narcissists can never have enough gratification. They are not *nice*—as Anna Freud would say—they do not value relationships with others of a different perspective. Even mature narcissists, like the old-boy networks that Louis Tiffany plumbed, treat their privilege as well deserved by exaggerating their benevolence toward women or the poor. But immature narcissists overtly claim a right to own women. Is that how raw aggression enters cultural regulation? Is bound aggression intrinsic to social development? Parents use bound aggression to put reality limits on their child's behavior. After reflecting on the raw, unbound aggression of World War I, Freud came up with a structural theory that put aggressive energy at the disposal of the superego. Anna reshaped his theory, by making aggression equal to libido in the economics of mental energy. As a woman she knew how much male cultural aggression limits a woman's role.

In *Group Psychology* (Freud, 1921), Freud described the instinctual role of aggression:

> In the undisguised antipathies and aversions which people feel toward strangers with whom they have to do we may recognize the expression of self-love—of narcissism. This self-love works for the preservation of the individual … as though the occurrence of any divergence from his own particular lines of development involved a criticism of them and a demand for their alteration. We do not know why such sensitiveness should have been directed to just those details of differentiation; but … in this whole connection men give evidence of a readiness for hatred, an aggressiveness, the source of which is unknown, and to which one is tempted to describe an elementary character. (p. 102)

Here, Freud discusses aggression as a social instinct in its own right. He knew that aggression was part of every human developmental stage, yet he wanted to see aggression as deriving from infant love/hate ambivalence. However, the antithesis lies between libidinal wishes and cultural regulations in each stage of development.

Each culture presents a unifying myth to its members. Psychoanalytic concepts were so popular in America that they played a major role in turning popular culture toward a mythic individual exceptionalism: a narcissistic culture. But mature narcissistic leaders must bind their aggression, because when they are too selfish, the disparity between their access to pleasure and that of others turns the public against them. For a long time American access to getting ahead—the American dream—moderated that cultural conflict. Hollywood played a role in popularizing psychoanalysis, using it to rationalize American mythology. During the depression, Hollywood sold the American dream, portraying men who successfully overcame their conflicts and challenges, and women devoted to a man who turned out to be wealthy and exceptional. Freud and Anna never trusted Hollywood—or movies—to enhance their cultural influence.

NEUROPSYCHOANALYTIC SOURCES OF CULTURAL CONFLICT

Let's return to Iain McGilchrist for his analysis of the transition to modernism in the early twentieth century, and to postmodernism in the approach and aftermath of World War II. McGilchrist sees twentieth-century cultural process as denying the RB's holistic and affective approach to the world, in favor of a LB, abstract approach honoring science and logic, for example, technology used to rationalize narcissistic power. How does narcissistic cultural leadership become tyrannical? Is it an excess of raw aggression, or a rationalization of a right to have more pleasure than others? I think collective narcissism among economic leaders of a culture leads to cultural regression, the unbinding of raw aggression among those wishing to remake the cultural world in their image.

The simple answer is that the 1 percent of the 1 percent of the people in the world who control the money that is the cultural price of survival are too self-centered to care if they destroy those who want an equal chance to survive among their loved ones. The more complex answer is that with a rupture from communities that have gone on for generations, with massive uprooting and migration of masses of people, with the nuclear family separated from its forbears, people lose touch with the economic organization of global culture. They are only aware of those who give them trickle down, even like drug dealers that supply the drugs needed to feel intact symbiosis.

The deeper answer resides in the effect of cultural dissonance on the biological sources of cultural identity. The hippocampus in mammals is the source of place and location, and the source of present and past. Long-term memory underpins individual and cultural identity. McGilchrist shows the importance of the hippocampus to place:

> The "homeless" mind: attachment to place runs deep in us. In neurological terms, the evolutionary roots of the integrated emotional system involved in social attachments may lie in more ancient and primitive animal attachments to place. Some animals bond as much with their nest sites as with their mothers. (McGilchrist, 2010, p. 390)

Nietzsche is McGilchrist's epitome of a twentieth-century LB consciousness pursuing self in place of others. Pure self-consciousness destroys itself, so McGilchrist is sorry for him:

> The more we rely on the left hemisphere alone, the more self-conscious we become; the intuitive, unconscious unspoken elements of experience are relatively discounted, and the interpreter begins to interpret—itself … Nietzsche's "speech about speech" … is a lonely self-enwrapped one: "nothing speaks to him anymore." (McGilchrist, 2010, p. 399)

McGilchrist describes a LB, virtual world of narcissistic isolation, alienation, fear, irony, and cynicism where words speak only to words at the expense of empathy. Freud realized that latency writes each person's self-awareness into terms of cultural identity, adding preoedipal family myths to postoedipal cultural identifications. Luria clarified this development from the vantage of higher cortical brain function:

> … higher forms of human mental activity are sociohistorical in origin. … man is born and lives in a world of objects created by … society and in a world of people with whom he forms … relationships … this milieu influences his

> mental process. The natural reflexes of the child (sucking, grasping, etc.) are radically reorganized as a result of the handing of objects. New motor patterns are formed, creating ... virtually a "mold" of these objects, so that the movements begin to match the properties of these objects. The same applies to human perception, formed under the direct influence of the objective world of things, themselves of social origin. ... (Luria, 1980, p. 30)

In conjunction with findings of Pavlov, Vygotsky, and Piaget, and examination of Russia's World War II brain-damaged veterans, Luria concluded that a *second signal system* using the inner speech of reflection socializes higher cortical prefrontal functions of the mind. It maintains mental functions even though their development depended on primary and secondary networks. Describing the same mental development, Freud realized that the comprehensive ego identity of latency is based on the capacity for continual verbal reflection on identity. Latency develops the capacity to repair identity—and to lie. It is only when reflection synthesizes each hemisphere's integrations that biological necessity and social reality provide a sense of reliable human identity.

DOES CULTURAL REFLECTION EVOLVE?

Freud thought that the development of narcissism as individuated mental life challenged the integrity of tribal culture by introducing human-made trauma. He concluded that parricide and infanticide so disrupted earlier forms of tribal structure that more hierarchical forms of culture evolved. He used subjective reflection on his life development: self-analysis, as a story that also reflected cultural evolution. Working together, leadership councils joined their reflections to develop and implement laws for cultural behavior. They sublimated their own narcissism for the sake of cultural progress. Like Luria, Freud concluded that cultural evolution led to biological evolution by further developing the human capacity for reflection. Thus, oedipal resolution led to cultural participation in a prolonged latency, an evolved mental development interposed before the hormonal changes of adolescence-reinforced instinctual development. That led to diverse cultural roles as individuals entered the structure of cultural responsibility. But a return of the culturally repressed narcissism with technological forms of mastery of others, such as better war weapons and convincing internet communication, led to war trauma that compromised the evolution that made us logical creatures.

Perhaps we are reentering a RB world view. Twentieth-century world wars and the regression of our present cultural hierarchy foster left brain overkill with words that speak to themselves in lies, like the exemplar Trump. *Make*

America Great Again was his immature narcissistic appeal to the collective phantasy life of those who feel they should be great and angry because the greatness has not reached their life. Trump is an immature narcissist because he doesn't have self-observing reflection ready to question his logic. Trump has the grandiosity that appeals to have-nots, but he does not distinguish reflective truth from lies. When a culture votes for grandiosity and gives aggression free rein, it regresses into cultural syndromes. When hard times hit home, the brain seeks archaic phantasy.

THE NEUROSCIENCE OF CULTURAL COMMUNICATION

Matthew Lieberman, who coined the term *Social Cognitive Neuroscience* for a science of sociocultural inferences based on neuroscience research studies how the brain regulates social attitudes through social codes. TPJ networks linked to DMPFC networks process social communication as we develop a *theory of mind* (ToM). The TPJ is where outcomes of social interactions are stored as successful or unsuccessful choices. Lieberman studied social attitudes by detecting activation in these brain networks when subjects recognize their own ToM. Like archaic tribes using tattoos as iconic signs of group membership, social language is recognized through *buzz phrases*. Our use of smart phones has even evolved buzz words and familiar icons into a virtual unified culture.

Lieberman studies social familiarity as the basis of empathy. Studying response to Democratic or Republican politicians, subjects show higher interest, for example *DMPFC-activation,* in their own party preference. Well-known politicians from both parties also trigger high activation. Choices of what is familiar and therefore satisfying in the past is part of what enters the global neural workspace consciously as we make our cultural decisions and choices. The DMPFC maintains both generic and person-specific social familiarity (Welborn and Lieberman, 2015). They show that ToM familiarity activates the TPJ in both members of a social dyad—adding buzz to seed ideas. American advertisers use their research to persuade target groups to commit to a product. O'Donnell, Falk, and Lieberman (2015) show that positive (feel good together) and negative empathy (your enemy is my enemy) are correlated when it comes to a dyad's TPJ activation. Subjects with high dyad empathy show low empathy toward strangers.

A study of whether the DMN readies a person to relate to another mind shows that an "intentional stance" is the prelude to expected communication (Spunt, Meyer, and Lieberman, 2015). The authors show that the medial DMPFC is the part of the DMN that sets readiness:

> We found evidence that variability in spontaneous dmPFC activity both within and across participants has a priming-like effect that is selective for mind-focused judgments of other people. We offer this as strong evidence that DMN activity in between moments of cognitive activity is the biological basis for the powerful human tendency to adopt the intentional stance. (Spunt, Meyer, and Lieberman, 2015, p. 7)

I conclude that a conscious network of the DMN sets the LB into an intention to judge communication, while other networks respond to difficulty in results requiring private fantasy thought about further interaction. The intentional stance seems equivalent to superego. Finding this function in early infancy and in chimpanzees, the authors conclude: "Evolution seems to have made a 'bet' that the best thing to do with any spare moment is to get ready to see the world in terms of other minds." (Spunt, Meyer, and Lieberman, 2015, p. 8)

Social cognitive neuroscientists know that language use can discern whether both parties are part of the same subculture. In our world of sensitivity to bias, *micro-aggressive* forms of speech are recognized in terms of whether they foment raw aggression or whether they are harmless. The human brain is organized for social communication within familiar groups of kin, tribes, religions, or political leanings. Social language within a given socioeconomic group is acknowledged through familiar language codes. Nation states require overlap among indigenous and migratory tongues. Without a patriotic myth, mixed culture can disintegrate into conflicting social tongues. Freud believed instinctual conflict triggered regression to preoedipal (archaic) stages of cultural development. Are love, aggression, ambivalence, and bias built into the brain, or are they conveyed through culture, or is it a combination of both?

Chapter 13

The Pretense of Cultural Leaders

Chapter 13 examines Freud's later life cultural works dealing with the effect of political leaders on individual experience. He experienced the Gestapo closing in on his family and shutting down psychoanalysis in Europe, as many of his psychoanalytic colleagues migrated to America, especially to New York, by 1933. As the last man standing for psychoanalysis, and a believer in the capacity of the collective unconscious to right itself, Carl Jung accepted the position of Germany's chief psychiatrist of analytical psychology. Freud's relationship with William Christian Bullitt continued as the two of them worked on their book, *Thomas Woodrow Wilson: A Psychological Study* in the late 1920s and early 1930s. Freud's discussions with Bullitt in the early thirties, until Bullitt was appointed by his friend Franklin Roosevelt as the first American Ambassador to the Soviet Union in 1933 was the reason why his book with Freud was withheld from publication. Freud had been as wary of the Bolshevik's experiment in national communism as he was of Germany's experiment in National Socialism, as he was of America's god-given national capitalism. He and Bullitt questioned how leaders of nations can be so pathologically self-centered, grandiose, and delusional. Wilson was their exemplar:

> Fools, visionaries, sufferers from delusions, neurotics and lunatics have played great roles at all times in the history of mankind ... it is often ... the pathological traits of their characters, the one-sidedness of their development, the abnormal strengthening of certain desires, the uncritical and unrestrained abandonment to a single aim, which gives them the power to drag others after them and to overcome the resistance of the world. (Freud and Bullitt, Introduction, 1966, p. xvi)

The above quote is from Freud's Introduction to their book. Chapter 13 investigates how immature narcissists become cultural leaders by offering illusions of how needs can be satisfied. Distinguishing mature from immature leaders, Freud decried those who promised well-being when reality could not provide it. Putting this in cultural developmental terms he reconstructed how in crises shamans brought group emotions to bear in carrying out the leader's decision. I cannot help comparing the populism of our current era led by President Trump, with the formative eras of tribal changes in culture that Freud described in *Totem and Taboo*. Journalists were the shamans of nineteenth- and twentieth-century culture, gathering information from influential leaders and putting it together emotionally as ongoing cultural narrative. Some shamans used group emotion to become leaders themselves. As Freud implies in the quote above, Wilson used his writing and speaking to claim a religious delusional identity as leader and religious shaman rolled into one.

William Christian Bullitt, born in 1891, co-author with Freud of the Wilson study, was a journalist of his times who impacted culture. As we follow the Freud/Bullitt analysis of Woodrow Wilson's policy during World War I peace negotiations, we'll explore Bullitt's role in dealing with Clemenceau, Lloyd George, Lenin, Wilson, and Colonel House. In his *Introduction*, Bullitt describes how their book came about. After seeing Freud for analytic work on his first marriage, Bullitt visited him in Berlin in 1922 as Freud was recovering from an operation for palatal cancer. He spoke of planning a book about Wilson; Freud asked to join in the writing, to show how one man's delusions affected Western culture.

At the time, Freud was trying to break his addiction to cigars that affected his heart and caused his cancer. He was further depressed from watching Heinele die from TB. Unlike himself, he felt, this brilliant child deserved to live. Freud's cultural works always alleviated his depressions triggered by losing self-objects. New work would give him a lease on life. Impressed with Bullitt's access to world power, he absorbed and tried to modify Bullitt's perspectives.

WILLIAM CHRISTIAN BULLITT'S BACKGROUND

Bullitt grew up in Philadelphia's Rittenhouse Square, a center of American aristocracy, where everyone who was anyone knew one another. An aristocrat with revolutionary spirit, he reveled in exposing the aristocratic persona. His father's wealth came from investing in Virginia coal mines, while providing legal counsel to the Norfolk and Western Railroad. Growing up close to his mother who was emotionally invested in his success, he traveled to Europe with her studying German and Russian. He graduated Phi Beta Kappa from Yale, where he was voted the most likely to succeed and was a popular member of the dramatic society with his friend Cole Porter. He began Harvard Law School, but when his father died in 1914, he dropped out. He, his younger brother, and mother had enough money to keep them in luxury for the rest of their lives.

William and mother went to Moscow, where he studied Russian until World War I broke out. Returning home, he worked for Philadelphia's *Public Ledger*. With a gossipy, biting style, his column became popular. Assigned to report on Henry Ford's initiative to end the war, he also got to know John Reed, whom he admired as a supporter of the Russian revolution. In 1916 he married Aimee Ernesta Drinker, a beautiful Philadelphia socialite, who preferred him to Felix Frankfurter. They set off for Germany where William was a war correspondent. In a summer tour through Germany, Aimee and William met the German intelligentsia, who freely shared their World War I views. In *An Uncensored Diary* of their trip, Aimee as Mrs. Bullitt gave a detailed view of the German response to America (Bullitt, Ernesta, 1917).

They found Germany so mobilized under Walter Rathenau that it seemed a matter of time before Germany prevailed. Rathenau told them that Zeppelin air power and a fast-expanding fleet of submarines gave Germany the capacity to mitigate British sea power. Germany had mobilized women to staff industrial production and to ensure that their children would still have good public education and necessary diet. Indeed, German education using Montessori methods was the precursor to Anna Freud's and Dorothy Burlingham's educational methods. Rathenau forewarned the Bullitts that Germany would use its submarines to attack American shipping if America favored and supplied the allies. During all of 1916 and 1917 until America entered the war on the Allies' side, Wilson said America would enter the war on the side that legitimately entered peace negotiations to end the war.

Bullitt returned as the Ledger's bureau chief. Eleanor and Franklin Roosevelt were part of the social circle in Washington that included the Bullitts, Walter Lippmann, Felix Frankfurter, Sumner Welles, Allan Dulles, and Colonel House, Wilson's confidant. Wilson was so taken with Bullitt that he asked him to make Bullitt his intelligence aide at the Peace Conference. Wilson asked Bullitt to help him set up the League of Nations.

BULLITT'S SECRET MISSION TO THE RUSSIAN REVOLUTIONARY GOVERNMENT

Bullitt's testimony to The U.S. Senate Committee on Foreign Relations, September 12, 1919, shows him in charge of intelligence briefing among the American delegates to the Peace Conference (Bullitt, 2004). During negotiations on the partition of Europe, the Bolsheviks were still at war in Russia's North, where British and American troops held Murmansk and Archangel. Lloyd George, House, and Bullitt realized that if the Bolsheviks agreed to stay out of the north, Britain and the United States could withdraw, simplifying all negotiations.

All that was still needed after a secret British meeting with Litvinov, a roving Bolshevik negotiator, was Lenin's withdrawal date so boundaries could be set up. Bullitt led a secret mission to Moscow that included Lincoln Steffens, a well-known leftist, to get a date for the withdrawal of allied and Russian troops to prevent mass starvation by bringing food supplies to the North. Meeting with Lenin, Bullitt reached an agreement for the United States to recognize the revolutionary government if it promised to confine itself to its present position. Despite Lloyd George's approval, France's Clemenceau tried to block it, arguing that negotiations with the Bolsheviks would legitimize them.

What was Bullitt's team up against with Clemenceau? By age 20 in 1861, Clemenceau was a throwback to Vendée revolutionary France, a member of a revolutionary brigade out to overthrow Napoleon III. Publishing a journal advocating demonstrations in the streets he was imprisoned for three months, after which he returned to medical school. Covering the American Civil War in 1865 as a correspondent, he not only learned how to influence the masses, also he acquired an American wife—which turned him against America. In 1889 he founded a radical newspaper, *La Justice*, using it to become a political leader. In 1898 his newspaper, *L'Aurore*, published Zola's defense of Dreyfus. In 1917, France was ready to capitulate to Germany's takeover of Paris, but Clemenceau, as Premier, rallied the nation to resist. By then he was strongly anti-communist, viewing the masses as producers of chaos. He was the hero who rescued France.

Wilson, House, and Bullitt were trying to reconcile the views of Clemenceau and Lloyd George, when a French anarchist shot Clemenceau—causing only flesh wounds. But French tabloids wrongly suspected a Bolshevik. The French fear of having their own Bolshevik revolution influenced the conference, ending any hope that Bolshevik Russia could be brought into negotiations. When Wilson caved in to Clemenceau, Bullitt was so disillusioned; he resigned in the following open letter in testimony to Congress:

> My Dear Mr. President: … I was one of the millions who trusted … implicitly in your leadership. … But our government has consented now to deliver the suffering people of the world to … a new century of war. … Russia "the acid test of good will" … has not even been understood … the present league of nations will be powerless to prevent these wars, and the United States will be involved in them by the obligations undertaken in the league and in the special understanding with France. … the duty of the Government of the United States … is to refuse … entering the league of nations, to refuse to entangle the United States further by the understanding with France. … if you had made your fight in the open … you would have been able to resist the pressure and … established the

new international order based upon broad and universal principles of right and justice of which you used to speak … May 17, 1919. (Bullitt, 2004)

Following Bullitt's testimony, the United States did not enter the League of Nations.

THE POLITICS OF AMBIVALENCE: FREUD AND BULLITT

Why did Bullitt and Freud wait so long to write the book and even longer to publish it? In Bullitt's *Introduction* he says Freud was intrigued that he and Wilson shared the birth year of 1856—he does not add that his father too was born in 1856. Perhaps in their cultural transference to each other, Freud felt Bullitt's ambivalence toward him, and ambivalent himself he did not want to put America (Bullitt) at the fulcrum of power. Yet, I can imagine how interested Freud was in all this inside information about America.

Bullitt's marriage to Aimee ended in 1923. She became a famous feminist and women's rights advocate. In 1924 he married Louise Bryant, the still grieving widow of the renowned American communist, John Reed. The 1981 film *Reds* documents Reed's and Bryant's journalistic coverage of the Bolshevik Revolution. Instead of pursuing the Wilson study, Bullitt vested his fervor in his new marriage by writing a novel dedicated to Louise entitled *It's Not Done*, which had a dozen reprints in its first year (Bullitt, 1926). A thinly disguised memoir, it blamed Wilson for failing to stand up to the French in the peace negotiations, after the journalist hero had negotiated a way to keep Russia in the European community. The novel also showed how the Philadelphia aristocracy could lose status as the poor became the nouveau rich.

Bullitt welcomed socialist thinking, believing like Freud that the lower classes would willingly become progressive if they had the opportunity to enter their economic culture with equal opportunity (Casella-Blackburn, 2004). Perhaps that also reflected his oedipal relationship of competition with his father. His socialist attitude, stronger after World War I, coincided with Freud's support of free clinics and nonmedical psychoanalysts. They both wanted to give the poor a share in cultural progress. Turning to their own careers, Freud and Bullitt postponed their book.

NARCISSISTIC INTRIGUE AMONG WORLD LEADERS

When highly narcissistic leaders meet to resolve world issues, the more immature ones may insist on having their way. As la crème de la crème,

Clemenceau allied himself with French narcissism. A working alliance among narcissistic leaders often builds intrigue. The group of journalists and public relations specialists who Wilson brought to the peace table impacted American world policy. Narcissistic journalists like Bullitt and Walter Lippmann, believed they controlled the information flow that induced cultural change, informing the public what they needed to know. Indeed, the United States maintained its tradition of truth telling, influential, journalists who brought the population into consonance with U.S. national policy, until false, alternative, news and overtly manipulative journalists and the media placed selling the news over reporting the vital news. That is a major sign of cultural regression in our own era.

Freud's followers did not realize how much Bullitt knew about the dynamics of the Allies in the Peace Conference. Instead, they were embroiled in the question of how much of the book was authentic Freud. Bullitt and Freud did agree, as manifest in the writings of each by 1930, that the Marxist experiment by the Bolshevik government of Russia posed a threat to world order; better they thought to bring them back into the European sphere of negotiated power politics. But Bullitt postponed the publication of the Wilson book when it was finished in 1932, because he had aspirations to influence the Bolsheviks to ally with America. Thus, he decided not to include new material that Freud wanted to add to the book. He believed that publication would disturb American policy toward Russia, when it still might be resolved through his efforts to influence Roosevelt.

After divorcing Louise, who became alcoholic and open about her female lover, and taking legal responsibility for their daughter, Bullitt was ready to return to public life. When Roosevelt ran for president in 1932, Bullitt wrote foreign policy speeches for him and was in touch with him daily on a special line. In 1933, he proposed to Missy, FDR's secretary. Bullitt always sought women close to leaders he loved—Roosevelt appointed him the first ambassador to the Soviet government. He had, after all parlayed his closeness with his mother to gain access to his father's money. But when Missy visited him in Moscow in 1934, she learned he had a lover—a ballerina; so she broke the engagement.

Given Missy's devoted relationship with Roosevelt, it is easy to surmise that Bullitt wanted to parlay his relationship with her into further closeness with Roosevelt. Many people in public life feel a narcissistic benefit in sexual relationships with the famous. Indeed, many cultural leaders have a sexual life with many partners—sometimes bisexual. Freud was envious of people like Bullitt who felt free to have lovers, yet he did not often analyze the secondary gain of libidinal bonding in politics. In the following recount of Bullitt's position with the Freud family and with the Roosevelts I am using many articles without citing them in order to make, but not defend my point. Freudians and historians have tended to dismiss Bullitt's position and the relevance of Freud's work with Bullitt. Political mythologies are legacies.

The Roosevelts had a closer ally than Bullitt in Sumner Wells, a descendant of wealthy, aristocratic, precolonial Americans. As a child, Sumner's emotional parenting was provided by black servants on his family's Connecticut estate. At age 12 in 1905, Sumner was a page boy at Eleanor and Franklin's wedding; at Groton he roomed with Eleanor's brother, and his cousin married Franklin's half brother. Graduating from Harvard, he followed Franklin's advice to join the Foreign Service. Specializing in Latin American affairs, he helped Franklin campaign for the presidency in 1932. In 1933, Franklin appointed him Assistant Secretary of State for Latin America. Then, as Undersecretary of State he was given even more responsibility than Secretary of State Cordell Hull for setting goals for American involvement in the coming World War II. Under the pressure, Welles drank heavily. In 1940, he was arrested for sexually soliciting black Pullman car porters. The Roosevelts protected him.

Despite becoming ambassador to France in 1936, where he was loved by the French, but not the Roosevelts, Bullitt was so jealous of Wells, so angry with Roosevelt, and so concerned with U.S. policy, that he went all out in exposing Wells' drunken episodes and homosexual behavior. He turned to J. Edgar Hoover to bring Wells down—Hoover had his finger on everyone's pulse, always suspicious of homosexual secrecy and communist involvement. But Roosevelt made sure Bullitt's political career was finished.

THE PRESIDENT WILSON CASE: SEXUAL IDENTITY DETERMINES CULTURAL ATTITUDES

Freud and Bullitt finished the Wilson book before Bullitt's falling out of grace. Examining Wilson's narcissism contributed to an understanding of world affairs and their irresolution. The Wilson Study is worth reading for those who want to understand cultural instincts, and how the resulting identifications form the dynamics of character. It shows how Wilson's sexual dynamics determined his political judgment and decision making. I think acting as Freud's psychoanalytic student in their study, Bullitt rewrote Freud's theories faithfully. They agreed on the effect of Wilson's dynamics on his politics. Bullitt's was the political mind that followed Wilson's decision making with disappointment throughout World War I and the Versailles Peace negotiations.

He knew firsthand how much Wilson loved Colonel House, depending on him almost completely to organize America's response to World War I; and

Bullitt felt loved by House. Bullitt told Freud how House found his way into Wilson's mind: He would give him advice with no response; two weeks later, he would give slightly different advice. Wilson would answer what House had said the first time as if his own opinion, and came to think of House as an echo of his own mind. Freud and Bullitt agreed: that is how to influence an overly narcissistic man.

Wilson idealized his father, a Scottish Presbyterian minister who constantly spoke of god's design, carefully training his son how to use words to convince others of the truth of religion. Young Tommy took the message to heart. Yet for his own individuation, Wilson made Gladstone, the British orator, an ego-ideal imago. Tommy could be greater than his father with a mind seemingly his own. Narcissists use their speech and writing to cunningly convince others of what to do without knowing their reality. Wilson lied about others' intentions to get people to side with him. If you were not for him, he hated you. Freud and Bullitt showed how libido leads to character formation:

> Repression is the method of denying the existence of the instinctive desire which demands satisfaction. … Identification seeks to satisfy the instinctive desire by transforming the ego itself into the desired object, so that the self represents both the desiring subject and the desired object. … Sublimation is the method of giving the instinctive desire a partial satisfaction by substituting for an unattainable object a related object which is not disapproved by the superego. … (Freud and Bullitt, 1966, p. 44)

Identification builds character by investing libido into objects that represent the self. Someone may love the child person he sees within the ego of the other as like himself. From his own childhood, Wilson always felt physically vulnerable: Anxiety affected his appetite and digestion, incapacitating him for long periods. He needed a wife to provide unwavering mother love. When his first wife died during his second term of office, he could not function until he fell deeply into symbiotic mother-love—then he was distracted from his duties.

WORK MAKES WORKS

What does a person leave behind in culture to show that he or she ever existed? The fantasy of being part of culture and adding to the myths of family, children, other kin, tribe, nation-state, religion, and science creates illusions that drive a person to keep working toward a culmination. For Wilson, the goal had always been adding to god's creation. Freud and Bullitt showed that his adolescent religious conversion experience transferred a large portion of his love for his father and his Christ-imago onto a boy a few years older, who planned to become a Presbyterian minister. Thus, at 16, Tommy loved a friend who converted him to make Christianity a life work. The Freud, Bullitt book shows how ego identity uses instinctual energy vested in ideal others as fuel for further identity passages.

Wilson's ambition to create a league of nations to show his life immortalized as Christ's had been, manifested his wish for permanent identity: not unlike psychotic identity formation. The wish to live and love forever is an aspect of symbiosis. Willingness to work for the benefit of others is considered mature identity, like parents who wish to live on through their children. Investing leaders with religious mythology provides a RB quality of lasting holistically and forever. Freud too turned toward cultural contributions more exclusively as he got older. Being president carries the godlike pose of working for your legacy as a permanent mark of identity. Historians often evaluate the influence of past leaders in terms of their own era. Freud knew that symbiotic love feels eternal.

ANNA FREUD COAXES BULLITT

Anna Moen Bullitt, his daughter with Louise, encouraged me to view his correspondence with the Freud family, which she donated to Yale University in 2008. Soon after Freud died, Anna Freud asked Bullitt to give her any parts of the book her father wrote in German, to show that those parts had been written by Freud. Bullitt refused, claiming the book could only be published as a whole—as Freud wanted. He added in a further letter:

> As your father may have told you, when I entered public life again it seemed to me undesirable to publish the volume, but it also seemed unfair for me to deprive your father of possible royalties and I therefore insisted on settling the matter by giving him a check for several thousand dollars. (November 28, 1939) (see Bullitt, Anne Moet, 2008)

In 1932 when Bullitt was working for Franklin Roosevelt's election he told Freud the book could not be published at that time; Freud did cash Bullitt's check. Not until 1966, after Wilson's second wife died, did Bullitt allow the book to be published—at Ernst Freud's request. (Ernst was adopted by Freud's family when his father, Max Halberstadt, was unable to care for him after Sophie died—because of his war injuries. Ernst changed his last name to Freud.) Bullitt wrote Ernst that he did not have Freud's German sections but he did have his Introduction written in German, which became Freud's introduction to the published book. Throughout the late 1930s, Bullitt used his position as Ambassador to France to intervene for the Freud family, whenever visas were requested by any family member. He used his diplomatic connections to get the documents for Freud and his entourage to immigrate to England, and even offered to help Freud come to America.

THE COEVOLUTION OF THE INDIVIDUAL WITH CULTURE

Freud and Bullitt showed the effect of Wilson's Christian delusion on an economically castrated Germany and an ostracized Bolshevik government. It is worth noting how much politics influenced Freud's thinking about *Civilization and Its Discontents* (1930), his clearest work on the relationship of the individual with culture. In that work Freud asks ironically if a fair apportioning of economic resources according to Marxist principles would give everyone an opportunity to pursue happiness without fear of harm from others:

> If the development of civilization has … similarity to the development of the individual and if it employs the same methods, may we not be justified in reaching the diagnosis that, under the influence of cultural urges, some … epochs of civilization—possibly the whole of mankind—have become "neurotic"? An analytic dissection of such neuroses might lead to therapeutic recommendations which could lay claim to great practical interest. I would not say that an attempt of this kind to carry psycho-analysis over to the cultural community was absurd or doomed to be fruitless. (Freud, 1930, p. 144)

Civilization and Its Discontents is a comprehensive reflection on the relationship between person and culture, so let's take a good look at it. If individuals worry about survival every day, while leaders take the spoils, they feel cheated and angry, yet guilty. Freud teases out the superego sources of aggression that became re-internalized after having been projected into cultural control in the past. Most people are willing to develop situational morality if their survival and their kin's survival seem protected. Thus, preoedipal anxiety of losing love and care because of aggressive behavior evokes *regret* at being caught and rationalized excuses, while postoedipal guilt uses the aggression that provoked the earlier fear of abandonment to *punish* the self for defying new cultural rules. Internalized cultural aggression directed against the self can lead toward suicide, or energize a wish to revolt.

Civilization and Its Discontents shows Freud's ideas about the evolution of the superego in culture—and its maintenance. Freud said leaders imprint their character on prevailing cultural morality. He emphasized that grandiose religiosity results in failed leadership. But the irreligious Marxism of the

Bolsheviks was just as narcissistic, putting the party before the people. Freud and Bullitt agreed that the Bolshevik government rationalized their destruction of the bourgeoisie to install their immature narcissistic brand of Marxism. Freud's theory of two contending basic instincts grounded his thoughts about both mental and cultural development. He concluded that the individual and culture are always at odds. After World War I, aggression appeared to overshadow libido, leading him to think another war could follow the same path. After discussion with Anna, he de-emphasized the biological sources of the mind, and emphasized cultural sources.

LOVE AND SYMBIOSIS

Freud began *Civilization and Its Discontents* (1930) questioning whether the oceanic feeling of oneness with nature can logically be equated with religious proof of a timeless entity. His friend, the writer Romain Rolland, uses the beauty of oceanic experience to ennoble his writing; but Freud says he cannot credit such a feeling as a universal way to equate self with reality. "It is a feeling of an indissoluble bond, of being one with the external world as a whole" (p. 63). Instead, he equates being in love with pathological states in which mental functions feel like they exist in the outside world, as they did in infancy. I will reference *Civilization and Its Discontents* through the end of this chapter.

He ties religion to symbiosis by way of infant helplessness needing protection. That protection evokes an imago that gradually becomes an externalized father-god. Without others to deliver pleasure and protection, life would be unendurable. As it is, he says, life is a staged series of painful emotional challenges that require self-adjustments to reality. But ego identity develops in layers, such that earlier stages coexist with later ones. At every stage, defenses against pain are vested in illusions. "Life … is too hard for us … too many pains … and impossible tasks" (p. 73). Mind develops by inhibiting the full affect load of pain, finding substitute satisfiers, and using substances from inside and outside the body (if they are available) that cancel pain biologically. A search for happiness uses compensation either as adequate pleasure or as relief from suffering.

What purpose do human beings serve? Freud treats the question as a religious one; he is more interested in what purpose each individual *pursues*. Characterizing people by their mode of seeking happiness, he shows that cultural status is the main organizer of their mental world. He lists ways of mastering suffering. Feeling rejected carries the most life pain for most people. If the external world depletes us we may renounce our instincts as

Yogas do. Those who renounce instinct the least, show the most instinctual pleasure. In sublimation, an alternative satisfaction turns life toward culture and away from instinct, and the pay-off comes in the pleasure that culture bestows when we enjoy the fruits of our collective minds. Freud writes of: "an artist's joy in creating, in giving his fantasies body, or a scientist's in solving problems …" (p. 79).

The pleasure of working toward cultural goals, such as reading or writing a book, is part of Freud's understanding of sublimation.

> … work … gives him a secure place in a portion of reality, in the human community. … Professional activity is a source of special satisfaction if it is a freely chosen one … it makes possible the use of … constitutionally reinforced instinctual impulses. … The great majority… only work under the stress of necessity. … (p. 80)

Identification with others in culture leads to contented identity. Using the male prototype as usual, Freud sums up male character types:

> The man who is predominantly erotic will give first preference to his emotional relationships to other people; the narcissistic man, who inclines to be self-sufficient, will seek his main satisfactions in his internal mental processes; the man of action will never give up the external world on which he can try out his strength. (p. 84)

Freud concludes almost ironically that the one aspect of suffering that comes under human control—culture—is seen as the main cause of human misery. He cites three socially conceived reasons: the low estimation of the value of human life by Christians; the belief that archaic cultures were happier than present cultures (the good old days); and his finding that contemporary cultures cause mental suffering and anxiety neuroses. He denies that scientific advances in communication, travel, and medical care enhance happiness. As I grow older, Freud's argument seems more and more true, the harder it is to adapt new technologies.

But he does think that scientific advances enhance the amount of narcissism in both individuals and culture. When human ingenuity establishes ways of surviving better and longer, Freud believes these accomplishments propel brain and cultural evolution. His prime example is that the capacity to control fire (bind aggression) brought men urethral pride in extinguishing fires, and women pride in keeping the hearth fires burning. Using tools as leverage to extend muscle power was another feat enhancing brains over brawn. Freud thought the extension of human survival with the aid of tools promoted hubris—feeling like a god. Projecting narcissism into culture provides a sense of helping people live together more safely. Reflection led to philosophy, religion, science, and technology; controlling nature led to animal husbandry and irrigated agriculture.

Freud concludes that culture strengthened with the realization that community is stronger than any individual. The community maintains rules for everyone so that laws apply to every member. What individuals relinquish instinctually will energize sublimated aims of the group, thus instinctual control evolves into group process. But the more hierarchical culture became, the more narcissistic was the life of leaders; they lost too much touch with their sense of family, tribal, and religious affiliations. They came to live too much in the LB as McGilchrist says about Nietzsche. The deeply conservative among them eschewed modernism for tribal conservatism.

FREUD'S CULTURAL CREDOS

Freud thought early Homo sapiens evolved upright to mitigate the instinctual stimulation from smelling the anal and sexual organs. Perhaps as Darwin thought, focusing on the sensory organs of the face increased the communication needed to deal with the environment. Prolonged thinking evolved through the brain's capacity to enhance energy production to maintain maximal delay before you leap. The leader's primal horde evolved in order to keep their erotic love object under control; while women stayed for their love of their children. Later in evolution Freud depicted archaic men working together out of mutual necessity for survival, settling in extended families, and then tribes. After the band of brothers killed the father to steal his erotic privilege, they had to develop tribal laws to prevent killing one another. Incest taboos provided rules that evolved through inhibitory mechanisms in the brain, mind, and culture. As cultures expanded hierarchically to maintain orderly restraint among individuals, administration became complex.

> The mode of life in common which is phylogenetically the older, and which is the only one that exists in childhood, will not let itself be superseded by the cultural mode of life which has been acquired later. Detaching himself from his family becomes the task that faces every young person, and society often helps him in the solution of it by means of puberty and initiation rites. (Freud, 1930, p. 103)

Implied in this view is the principle that cultural conservation through evolutionary layers of development continues as regressive wishfulness throughout everyone's cultural life.

IF YOU CAN'T BEAT THEM JOIN THEM

Freud took issue with the doctrine "love thy neighbor as thyself." Sensing that strangers wish us harm and disadvantage, as we do them, is always present. Freud agrees that private property is the focus of enmity between those who have and those who do not, just as it is for young children; but the amount of aggression necessary to join them is another story; Jews were not allowed property, yet they were surrogates for others' aggression: "... massacres of the Jews in the Middle Ages did not suffice to make that period more peaceful and secure for their Christian fellows" (Freud, 1930, p. 114). He adds that the Nazi dream of world domination held anti-Semitism as its complement, just as the Russian Communist persecution of the bourgeoisie did. He cites America as the land of narcissism, where leaders exist beyond civilizing rules.

Freud apologizes for his failure to have seen Thanatos (aggression) as an instinct of destruction opposed to what Eros (libido) brings together as life. He only wished he could believe in Jung's idea of one instinct—perhaps now espoused by Panksepp. He compared disregard of ambivalence to a nursemaid's story of heaven. At least *his* nursemaid provided an image of hell's fires. Yet, Freud remained confused about aggression used to gain selfish pleasure, from aggression during trauma: fight, flight, or paralysis. But does he treat instinct as biological, mental, or cultural? How can culture be part of social reality when it is collectively projected from the physiological core of all the people?

If the RB binds aggression and the LB binds libido, higher cortical functions may be said to judge what Freud considers instinctual conflict. He says the superego uses ego-bound aggression derived from early life experience to develop the feeling of guilt, used for cultural control. If endogenous trauma during developmental stage changes is universal, then trauma during stage changes sets up mental process to be wary of further trauma. Culture uses a person's innate aggression to fuel a guilty conscience. The anxiety about a loss of love and protection when the young child feels angry and defiant toward the one who loves him, as well as the deeper fear of obliterated self, persists throughout life. To preserve the love relationship, regret springs up in the aftermath of disobedience—regret at being caught. But when superego is internalized during latency, it is not just a fear of losing love, but conscience about an imagined aggressive act. That is how culture manages the brain.

Freud asks why the most saintly feel the most guilt. And why does misfortune evoke guilt? He answers that bad luck feel like the protector must have stopped loving him or her. God is teaching the individual a lesson: fate is divine will. In *Civilization and Its Discontents* Freud refers to Aichhorn's *Wayward Youth* in a footnote to explain how this happens:

> The unduly lenient and indulgent father is the cause of the child's forming an over-severe super-ego, because, under the impression of the love that they

> receive, they have no other outlet for their aggressiveness but turning it inwards. In delinquent children, who have been brought up without love, the tension between ego and superego is lacking, and the whole of aggressiveness can be directed outwards. (Freud, 1930, p. 130, Footnote 2)

Distinguishing bound from raw aggression, Freud adds that primal fathers were angry gods. He wonders if raw instinct was inherited as both unmitigated love and aggression in primal life—his father and the one-eyed-Christian were two versions of the leader. Even if splitting father imagoes was Freud's experience, his oedipal explanation rings true to him. His answer is always the inheritance of patricidal wishes: unknowable guilt.

> This remorse was the result of primordial ambivalence of feeling toward the father. His sons hated him, but they loved him, too. After their hatred had been satisfied by their act of aggression, their love came to the fore in their remorse. … It set up the super-ego by identification with the father; it gave that agency the father's power, as though as a punishment for the deed of aggression they had carried out against him, and it created the restrictions which were intended to prevent a repetition of the deed. (Freud, 1930, p. 132)

IN THE THRALL OF CULTURE'S UNCONSCIOUS GUILT

In concluding *Civilization and Its Discontents* Freud explores why cultures produce excess aggression, making the superego overly cruel, working against the happiness of most people. He attributes unconscious guilt to culture, blaming it for teaching children to deny the aggressive side of their human nature, just as it hides the facts of sexual life. He notes that many analysands are reluctant to get better, to improve their relationships—even after they understand how their emotional conflicts held them back. Declaring that the negative therapeutic reaction is culturally reinforced, he fully acknowledges the structural role of aggression from brain to culture to mind.

Where does the feeling come from that culture can read your mind? How does conscience become conscious? How does the superego know your secret disobedience?

> The fear of this critical agency (a fear which is at the bottom of the whole relationship), the need for punishment, is an instinctual manifestation on the part of the ego, which has become masochistic under the influence of a sadistic super-ego; it is a portion … of the instinct toward internal destruction present in the ego, employed by forming an erotic attachment to the super-ego. (Freud, 1930, p. 136)

The ego's warning anxiety is evoked under the watchful inner eye of reflection that regulates the mind, sets goals, and avoids mistakes, binding aggression with guilt. From an evolutionary perspective, the most anterior prefrontal cortex is the top-down regulator of mental life, assessing how the self is doing in terms of finding pleasure and happiness and avoiding emotional pain. But the moral quality of culture is situational, subject to changes in group psychology and vulnerable to the media. It is not too hard for powerful narcissists to take hold of the fruits of culture by lying about social causes. Prolonged cultural dissonance between the morality proclaimed by leaders and the religious or tribal morality of the people gives rise to cultural regression and cultural syndromes.

In Freud's version of how culture acquired control of aggression, men identified with their leaders' qualities of aggression. After the brothers murdered their father, their enmity toward one another led to idealization of the dead leader. Since he was no longer directly feared, his imago became the mythical core of a narrative of cultural practices to restore order. But, in coordinating group social life, the leader and the shamans must avoid letting frustration evoke too much aggression, which makes matters worse. Conceiving that any kind of frustration gives rise to guilt, Freud writes: "When an instinctual trend undergoes repression, its libidinal elements are turned into symptoms, and its aggressive elements into a sense of guilt" (p. 139). His formula fits brain regulation; the LB promotes pleasure, the RB caution.

Freud concludes that individuals seek happiness, while culture seeks restrictions. Yet each community has its own superego "based on the impression left behind by the personalities of great leaders" (p. 141). He characterizes many leaders as monomaniacs who emphasize a unique perspective on cultural life. They are often reviled or persecuted in their lifetime and mythologized after death—like Jesus. They may feel mistreated by culture, and may become eccentrics and revolutionaries—like *Don Quixote*—a salient persona for Freud from his early adolescence. A great man comes to represent a cultural superego once his imago becomes iconic as a morality for the community—like Moses.

Of course, the problem is that different cultural communities have disparate moralities and founding heroes, with the resulting hostility leading to cultural conflicts. Cultures, like groups, cannot be as rational as individual reflection in understanding the sources of conflict. Freud points out that cultural morality requires everyone to obey its laws despite individual circumstances. He adds that religious proscriptions based on denying aggression, produce an ethic most people cannot follow when their survival is threatened. He wonders if cultures undergo their own neurosis, their own syndromes:

> [regarding] the therapeutic application of our knowledge, what would be the use of the most correct analysis of social neuroses, since no one possesses authority to impose such a therapy upon the group? But in spite of all these difficulties, we may expect that one day someone will venture to embark upon a pathology of cultural communities. (p. 144)

Who would be so foolhardy?

Chapter 14

The Evolution of Fantasy

Chapter 13 left off with Freud's use of instinct as a biological determinant of mental and cultural conflict. We have seen that the problem for making culture less onerous and aggressive, and to get agreement about how to regulate culture in a fair way, requires dealing with innate self-centered wishfulness. How can we turn fantasy into useful creative policy that enhances culture? Perhaps that is not possible without educating ourselves about how the mechanisms of fantasy work in the brain, how they become a useful part of our emotional life, and how they enter the regulations of the law within cultures. Chapter 14 looks to the DMN fantasy mode of thinking, as the switch mechanism among the three levels of organization: biological, mental, and cultural. Given unusual experience, emotional responses trigger the consolidation of new learning into the brain's plasticity. During fear, mental identity is disturbed by shifts in phenotypic responses to experience. Chapter 14 further explores the transition among these structural organizations, using contemporary neuroscience findings. Let's turn to Marcus Raichle to explain the brain's high energy use during its DMN core function as the predictive instrument of the mind through the consolidation of new learning.

> The idea of the brain's dark energy had its origins in my thinking when the large difference between the energy devoted to behaviorally evoked activity (very small) and that devoted to ongoing brain function (very large) became apparent to me. … I wanted to highlight the neglect that this discrepancy has endured and the lack of understanding that has resulted. … I have unashamedly borrowed the term dark energy from cosmology … by calling attention to a major constituent of the universe (i.e. energy) that we have yet to fully understand … it seemed to me that dark energy applied equally well to the brain in terms of its ongoing energy consumption. (Marcus Raichle, 2015, p. 26)

Raichle shows how the brain's predictive function depends on its intrinsic work, its readiness to move from one network to another, when many networks must participate in identity change. Fantasy is always about identity change; it prepares us to deal with the problems of survival. Raichle realized that the need for new learning to solve difficult problems of survival activates the DMN so completely that its intrinsic function produces a whole ionic shift second by second, as astroglia transfer energy to dendrites in the need for synaptic shifts. Priming some intrinsic networks forms new priorities for shifting consciousness among contributing networks in the brain's global workspace. I have wondered how the connection between diverse brain areas can be transcended when there is no direct axonal linkage from one network to another. In particular, I wondered how our medial anterior frontal workspace that manages our identity consciousness could make use of our hippocampal reception of new cognitive and emotional information, when there is no direct axonal linkage among their networks. Raichle found that slow wave energy is produced by the brain's heavy ionization of dendrites, preparing them for predictive work. The change in ongoing structure is a prelude to their new predictive priority among the networks of familiar consciousness that have psychological and cultural significance.

Raichle describes the high-energy requirements of the large brains of Homo sapiens, with their tremendous number of neurons and supporting glia. Much of the body's energy is expended on the DMN, which triggers daily updates of the brain's capacity for social relations and communication. Since the brain is a predictive instrument, daily change in social and cultural relations is vital for adapting to a changing milieu. Social problems emerging every day require physical changes in the brain's linkages, as the DMN consumes most of the brain's energy.

Raichle's study of energy waves during functional brain imaging clarifies understanding of the mind/brain intersection. The problem-solving part of the brain's work, such as salience detection and executive control, use blood oxygen level dependent (BOLD) signals of oxidative metabolism, showing up in flairs of activated brain imaging. But the bulk of the brain's oxidation shows up in very slow waves, which researchers mistakenly treated as irrelevant noise (Raichle, 2015). The evolution of mammals required a consistent energy source for their larger brains' intrinsic functions, as well as a ready source for practical thinking.

This chapter explores the DMN's fantasy-state functions as intrinsic neural organizers of the social brain. Medial networks elaborate the relational mind both as the core of human brain process and as the basis of social identity that identity builds up from conflicting experiences of love and aggression: distinguishing kin from stranger, friend from foe. Fantasy tracks relational changes; much of our mental life devolves on learning how to change our expectations of others, of our own self, and of culture (Tacikowski, Berger, and Ehrsson, 2017). The researchers cited showed that the medial prefrontal anterior cortex manages the intrinsic consciousness of identity as part of its DMN function. Our survival depends on how we process self–other relationships.

Freud's concept of brain/mind evolution expresses Darwin's observations of how animal communication supports survival by creating animal culture. And neuroscientists interested in human evolution are making scientific discoveries that provide context for Freud's findings. Speaking at the New York Psychoanalytic Institute on May 2, 2015, Panksepp discussed the discovery of a gene for brain enlargement in hominids that allowed a massive increase in neurons for human learning. Von Economo neurons have evolved as short-cut links among prefrontal limbic and neocortical networks, to provide quick reads of emotional factors that affect social decision making. And through the global social/cultural internet, human social signals spread even further and faster.

Raichle shows that the DMN mediates between immediate and future-oriented problem solving. In fact, the brain's intrinsic organization uses most of its energy in the changes required for predictive performance. As a hub of medial cortical networks, the DMN mediates between cognitive goal formation, executive function, reality-salience, and the intrinsic functions of the brain's daily renovation. Whatever seems emotionally unexpected is an occasion for the DMN to trigger change in medial network nodes of memory storage. The posterior cingulate cortex (PCC) is the highly active metabolic centerpiece of the DMN, prioritizing consciousness in response to new learning from the previous day. The reflective, dorsal part of the medial prefrontal cortex, functioning as a superego and observing ego, judges whether it is possible to carry out productive problem solving, or whether behavior must be inhibited while new priorities are determined by the PCC's linkage to past experience. The superego signals a change is necessary in moral identity standards. The ventral medial prefrontal cortex (VMPFC) processes signals of need, anxiety, and danger as part of both the DMN and the executive control networks.

Stephen Porges (see chapter 4) showed how the midbrain autonomic system in mammalian brains evolved to maintain prolonged high energy consumption. During difficult problem solving, limbic cortex triggers the autonomic nervous system to keep the brain steady in its mode of mental processing by switching to high attention and concentration, in which a solution can trigger a consolidation of new learning. But stressful situations trigger vigilance responses instead, leading to emergency affects that cannot be managed by new learning alone. And these require extensive brain change, which cannot occur within a day.

Linked to drive sources like dopamine and norepinephrine, the amygdalae (one in each hemisphere) are nuclear centers of the limbic system, outputting to other centers of emotional regulation. They pass signs of imminent threat to the hippocampus. During vigilance, the right amygdala activates the hippocampus to process its evoked emotional triggers by associating them with past emotional episodes. But with strong fear signals, the right amygdala blocks hippocampal reception, which reduces the prefrontal capacity for problem solving. In trauma, the brain's intrinsic functions undergo a chronic change.

LINKING FANTASY TO INTRINSIC BRAIN FUNCTION

Executive functions carry out necessary practical thought using oxidative energy. When practical thought doesn't work out in a familiar way, emotional, cognitive, or both may be the reason. So the DMN interacts with executive functions to mediate novel experience to promote identity integration and gradually learn how to carry out necessary problem solving. Cognitive learning goes on with less complication than emotional learning, possibly due to simple changes among neuronal synaptic linkages. Emotional learning is complex and cannot be consolidated in a day; cultural and social changes involve a complex reorganization of identity. As we'll see in more detail, slow wave sleep (SWS) consolidates fresh cognitive learning, while REM dreaming tries out emotional learning, requiring a more energetic process of change that links to many medial networks involved in emotional behavior. There is an extensive Freudian literature about REM dreaming as a core of fantasy life where self and object images freely exchange who is acting.

Thinking about your situation in life, you may feel narcissistic confidence about predicting likely successful social interactions. But frustrated, you may turn to new fantasies requiring brain revisions. The DMN activates more highly in the LB; while the salience and executive networks, which respond to the environment with attention that modifies behavior, activate more highly in the RB. What appears to an observer as a resting state is actually an energized state of intrinsic activity in neurons biologically changing with experience. Perhaps REM dreaming foreshadows intrinsic brain function change, manifesting mental image wishes produced by previous day experience. As the night progresses, REM periods get longer, perhaps indicating changes in DMNs as they prepare for the coming day. Mammals spend more time in REM sleep early in postnatal life when new learning and brain plasticity are high.

Brain imaging during SWS and REM sleep shows separate forms of consolidation of experience, distinguishing between practical thinking and DMN changes in response to unexpected experience (Dang-Vu, 2010). SWS studies show consolidation of new learning consonant with practical thinking; whereas response to threatening experience is only partially consolidated during REM sleep, which may take many days to complete. SWS deprivation interferes with learning consolidation; REM deprivation leads to disturbance of adaptation.

SWS deactivates the VMPFC, detaching consciousness of need; but REM sleep activates the VMPFC and the hippocampus, manifesting needful wishing. I think REM sleep allows us to imagine how to adapt our identity to major changes in the social and cultural milieu. SWS consolidates new declarative memory, for example how we verbally track altered practical activity, that is, planning; whereas REM sleep deals with protracted changes in identity, investigating how past wishes were fulfilled. Think of the collective feelings about cultural breakdown during the presidential campaign in 2016, and its aftermath in 2017 when people had anxiety dreams about their American dream, preparing for moral change in their future.

THE MIND IN CULTURE

> Evidence suggests that the DMN is a basic and phylogenetically old feature of human cortical function: The basic elements of the DMN can be observed in human neonates ... the DMN and its activity in between moments of directed thought may be evolution's solution to the problem of other minds. (Spunt, Meyer, and Lieberman, 2015, p. 1123)

Here, the authors suggest that the DMN evolved as a response to life in the social world. Does otherness exists before a sense of individual self? The part of the DMN that is fully conscious is directed toward social communication, while the part that stays preconscious mediates between relationships and daydreams. The authors conclude that the DMPFC linking to the medial temporal lobe within the DMN equates one's own mind with another mind in the *intentional stance* of a virtual dyad. In my mind that defines fantasy. Is infant intentional stance a nascent function of fantasy? Think of how the baby puts food in the mother's mouth with so much pleasure. The authors relate intentional stance to "priming": associating other's needs with one's own: "DMN during periods of rest might serve as an endogenous prime that makes an intentional stance the default strategy for making sense of the social world" (p. 1117).

If the DMN uses fantasy to regulate social communication, we may think of fantasy as a way of balancing metabolic needs with social conditions. The DMN's LB readiness to fulfill needs socially links to the executive function more laterally in the LB. Using inner speech to carry out intentions; our words follow our intentions into brain networks as preconscious language (Swanson, Eichele, and Calhoun, 2011). But if you think about what it is like to emerge from a daydream back into practical work, you may realize that a few conflicted words from the fantasy state remain in your mind. They would have stayed a jumble unless you happened to catch them at the moment.

How can a person still have a sense of identity when his medial temporal lobes including the hippocampus has been destroyed—like the famous patient, H.M.—erasing the events and episodes responsible for his sense of identity? Despite those losses, there is a set of networks preserving the human identity sense in the temporal parietal junction (TPJ) which records the usual outcomes of interactions with different types of people. These networks have recorded behavioral choices and their outcomes in situations of emotional communication during a person's lifetime. The TPJ is the source of what Matthew Lieberman calls theory of mind (ToM). He views social familiarity as the basis of collective empathy. His research focuses on nodes of the DMN that maintain the ToM, which includes the VMPFC that we have been equating with self. Quandaries about identity are part of intrinsic mind-wandering, when immediate, goal-oriented process is not engaged. In other words, fantasy thought about self and other is part of the DMN's intrinsic function. The trick is how to get others to agree with you.

The intentional stance manifests mental relationships in a dyad. Familiarity, like kinship, is an intersubjective experience of merging our DMNs—as Freud often said, I assume my mind is your mind. Lieberman uses these perspectives to elaborate the nature of public relations and advertising, which shows how dyad identity enters popular culture. ToM familiarity activates the medial temporal lobe in both people of a dyad (Meyer et al. and Lieberman, 2015). We prepare to greet others as familiar or strange within 0.2 seconds, well before we can bring lessons against bias into play.

Psychopathology based on trauma ensues when the DMN cannot promote intrapsychic adaptive change in identity. Research in the aftermath of trauma shows changes in DMNs. A Chinese group investigated functional mental networks of people two years after they survived a severe earthquake. Changes in their DMN coincided with PTSD development (Du, Ming-Ying et al., 2015). Trauma leading to PTSD triggers DMN reversion to past modes of social maturation—mental regression.

FANTASY DEVELOPS WITHIN THE MIND'S SENSE OF CULTURE

As people look back on their life, their childhood wishes, fantasies, and disappointments return, embedded in a familiar culture that has become past history. As Freud aged, he reflected on the stages of his life, his patients' lives, and on culture from archaic times. We have seen how the reemergence of his infant imagoes reshaped the transference in his relationships, and how he always mitigated the pain of losing self-extending others by constructing an oedipal narrative to subdue his anxiety about the continuity of his identity. Creating a cultural movement was another way he sublimated lost love.

Freud's cultural reconstructions gathered his infant prototypes of identity into personal myths—the DMN's revenant sources of himself. It's no wonder that Freud looked to fantasy as the libidinal gathering source of human and cultural creativity, since imagoes imbedded in fantasy are diffuse, powerful, and symbiotic, suitable to alter the imagery of forebears. As his identity-splitting became more severe with illness and old age, Freud reconstructed his imago of Moses as a non-Jew—the father of the primal horde—an attempt to resynthesize his own role as leader of his movement's cultural identity (Freud, 1939). When a foundational myth develops, the raw aggression of forebears is transformed into a story of overcoming aggression with love. In reconstructing his patients' past lives, Freud reflected on their identity sources.

SCIENCE MAKES THE NATION: HUBRIS BREAKS IT DOWN

One culture can enslave or destroy another with better weapons, strategies, or luck. But imposing new morality in the conquered culture does not change the balance of love for kin or aggression toward strangers. Living in a milieu of diverse cultural groups improves the chances for overall cultural progress, since the cooperation of many individuals from different perspectives enhances logical thinking. The resulting diversity leads to more individualism. When diverse tribes and religions take part in a hierarchical culture for economic reasons, they develop allegiance to the state or nation. Without allegiance, or with a feeling of being discriminated against, conditions are rife for the development of more narcissistic sociopathy in those cultures. Immature narcissists in long suffering cultures come to match the degree of immature narcissism in the prevailing culture.

Ever since rationalized war became the modus operandi for colonial power, narcissistic leaders have dominated the moral high ground in economic rule-making. It seems that the higher up you go in the economic scale of control, the more widespread empathic morality is left behind. Learning the secrets of those in power leads to cultural intelligence that nineteenth- and twentieth-century journalists and writers used to gain influence. The Western world they lived in featured a nineteenth-century romantic fantasy of human progress based on innovative scientific achievement used to justify cultural grandiosity. At the start of World War I, limited warfare was still an economic instrument for both the Allies and the Central Powers, a way to own the world through colonialism, rationalized as the right of a higher culture. Let's call this a collective phantasy of overly narcissistic leaders.

Our contemporary world has regressed even further into a hubris that excludes moral binding of aggression for the sake of personal power. Maldistribution of wealth, like the colonialism of the early twentieth century, causes cultural regression from the top-down; such that many people have discarded patriotism in favor of tribal and religious roots. Distorted economic equality results in the erosion of personal identity—it splits into discordant fragments from the holistic top-down. Widespread unbound aggression, for example immorality, leads a critical mass of people into traumatic sociopathy, depression, or psychosis. When nations regress into the chaos of terrorism and criminality, the result is a natural process of cultural destruction.

THE MIND IN PROBLEM-SOLVING OR TRAUMA MODE

People can use fantasy in creative ways to expand social and cultural consciousness and their own capacity for a maturing identity, or they can use it to deny a traumatic situation without developing strategies for the future. When we think of mind we usually think of its problem-solving mode of feeling and thinking, as if mental life were a puzzle about where to add each piece of information to make a realistic context for decisions. But when cultural, social, and physical realities threaten our survival with unrelenting stress, we move from ordinary problem-solving to a fantasy state, then to a traumatized state, where thinking becomes perplexed and feelings lose relevance.

Freud's *universals*, the developmental trauma of passages, tend to shatter infant symbiosis, oedipal fantasies, adolescent eternal love, and adult belief in eternal life. His theory that ego identity recapitulates stages of cultural development was fundamental to his understanding of how libidinal and aggressive development results in conflict. Laying the groundwork for syndrome formation, each stage affects fantasy life, sexual identity, character formation, and cultural participation. Each stage's way of binding love and aggression with defenses, triggered by anxiety or fear, also has its parallel in culture. For instance, infant love versus hate fuels tribal symbiosis; and young childhood willful power versus abandonment anxiety fuels religion. Collective phantasies of each stage persist in myths, religions, and cultural organizations. Based on these underlying cultural vulnerabilities, exogenous trauma (including organicity) can trigger individual and cultural syndromes.

FANTASIES OF MIND READING

The intrinsic split between reflection by the DMPFC and subjective identity feelings processed by the VMPFC gives rise to experiential phenomena over the life span—like having an indwelling mind reader. As I have learned from my own experience, passage into old age often simplifies a lifetime of reflections on what it means to have your own human identity and to pass it on. Aging with wisdom, people may imagine their legacies represent cultural value, but they also tend toward experiencing a second childhood. Early fantasy-wrapped imagoes may prompt a repetition of emerging identity, split from old-age identity by cognitive decline.

The DMPFC maintains the cognitive otherness of reflection by its linkage to conditioned dorsolateral prefrontal cortex. When organically induced decline in old-age cognition dissociates VMPFC self from DMPFC reflection, reflection may feel like a judgmental other from earlier life—identity reflected from past stages. A demented person mumbling in bed articulates the otherness of reflection as if it were another person speaking—like the otherness we can image exists in infant symbiosis (Harris, 2013). Let's consider Freud's aging reference to telepathy in his *New Introductory Lectures on Psychoanalysis* in 1932 [1933]. Those public lectures are perhaps Freud's best explanation of the cultural function of the superego. Because Freud had his own ego split, separating his mature reflection from a sense of a double, this could be projected into others as a transference relationship. Freud considered telepathy as a path for deep creative regression. He played telepathy games with Anna, whom he believed telepathic, and Ferenczi.

Goethe believed in telepathic communication with forebears, which influenced Freud's writing. In *Lecture XXX*, on *Dreams and Occultism,* he devoted 24 pages to debunking telepathy as wishful thinking, only to conclude with a paragraph opening the gates to it. By 1932, Anna was her father's mouthpiece; he could not articulate well because of his cancer. I think she replaced Julius as Freud's double by seemingly reading his thoughts. Dorothy Burlingham's work on telepathic twins and mother/child dyads added to Freud's feeling of reconstituting his family. He always used a double to represent his immersion into creative fantasy:

> Dorothy Burlingham, a trustworthy witness, in a paper on child analysis and the mother published ... observations which, if they can be confirmed, would ... put an end to ... doubts on the reality of thought-transference ... a mother and child are simultaneously in analysis, and reported some remarkable events ... the mother spoke during her analytic session of a gold coin that had played a particular part in one of the scenes of her childhood. Immediately afterwards ... her little boy, about ten years old, came to her room and brought

> her a gold coin which he asked her to keep for him. She asked him in astonishment where he had got it ... his birthday had been several months earlier ... there was no reason why the child should have remembered the gold coin precisely then ... the action had forced its way that day into the child's life like a foreign body. A few weeks later the mother was sitting at her writing-desk to write down, as she had been told to do, an account of the experience, when in came the boy and asked for the gold coin back, as he wanted to take it with him to show in his analytic session. ... the child's analyst could discover no explanation. ... (Freud, *Lecture XXX,* 1933, p. 56)

Freud had been analyzing Dorothy while Anna analyzed Dorothy's son, Bob. Freud must have made that association as he referenced Dorothy's analytic graduation paper. Indeed, he gave her a golden ring as a token that meant she had become a close disciple. He says here that if she is right, mental exchange between mother and child can be telepathic. Psychoanalysis brings the analysand and the analyst so close together than when the defenses have been analyzed, interpretation arises in both parties together, just as being in love feels like two persons having the same thought in the same moment. I trust that is not the result of the transference of high ionization from one brain to another in situations of high stress and closeness.

Freud believed that telepathy is a personal projection of empathy, a fantasy of closeness so extreme that mentalization occurs. Similarly, when many of the analysts whose lives we have visited tell their stories in the guise of patients, as Sigmund, Anna, and Dorothy did, they reveal their fantasy of wishful identification with others, as they move toward a change in identity to have their professional wishes realized.

FREUD'S LATE-LIFE MIND-BRAIN THEORY

Although Freud's structural theory corresponds to intrinsic functional neural systems, the older he got, the more culture beckoned him to widen his theory beyond the biological brain. Given what he knew about LB speech and RB assessment of reality; about anxiety regulation of behavior; about superego-triggered anxiety, guilt, and affect; about the ego as a container of life memories; and about fantasy wishes throughout all stages; his earliest neural theories converged with his later cultural theories.

The Dissection of the Psychical Personality in his *New Introductory Lectures* is his culminating presentation of the functional relationship among id, ego, and superego in human identity formation (Freud, 1932). Written when he was 76 years old, this work brings cultural influences on development into consonance with biological changes in the brain. Let's follow his conclusions in *Lecture XXXI*:

> From the very first we have said that human beings fall ill of a conflict between the claims of instinctual life and the resistance which arises within them against it: and not for a moment have we forgotten this resisting, repelling, repressing agency, which we thought of as equipped with its special forces, the ego-instincts. … (p. 57)

But culture itself reinforces and even structures ego identity, impacting both ego and superego. He asks how the ego as subject can take itself as object, and answers that it "can take itself as an object, can treat itself like other objects, can observe itself, criticize itself … one part of the ego is setting itself over against the rest. So the ego can be split. … Its parts can come together again afterwards" (p. 58).

Freud observed his own ego splitting and resynthesizing in the logic of secondary process. He knew splitting was horizontal—corresponding to how the two hemispheres of the brain experience the ego differently—as well as vertical—top-down versus bottom-up.

He was aware of his self-observing function, a critical agency in his mind feeling like an external agency, which viewed him and all his thoughts, reflectively and judgmentally:

> I formed the idea that the separation of the observing agency from the rest of the ego might be a regular feature of the ego's structure, that idea never left me … delusions of being observed … suggests that the observing is only a preparation for judging … we accordingly guess that another function of this agency must be … conscience. (p. 59)

Because love and aggression seem omnipresent, both in the mind and in response to others and to culture itself, Freud concluded that these two forces were the sources of the instincts of life and death. Thinking of melancholia, in which the superego attacks the ego, Freud realized that the culture-sourced superego can also approve of the ego, providing a sense of power in identity. He explores how feelings of moral anxiety develop in stages, for example in the anal stage, when morality is anxiety of being caught and so losing the love of the parent. Only in the postoedipal period does the superego become identified with prevailing culture beyond the family. The identification with parents, which gives rise to a superego observing the child from within, is subject to alteration by outside culture including peers and educators.

HOW CULTURAL IDENTIFICATIONS WORK

Identification with caretakers from early infancy on brings the qualities of others into the developing ego and the archaic superego. Freud adds in

Lecture XXXI that after the oedipal complex is resolved and the superego synthesizes both parents' qualities, further identificatory processes will reshape moral understanding—but not the superego. When

> the Oedipus complex gives place to the super-ego they [the parents] are something quite magnificent; but later they lose much of this. Identifications then come about with … later parents as well, and indeed they regularly make important contributions to the formation of character; but … they only affect the ego, they no longer influence the superego, which has been determined by the earlier parental imagos. (Freud, 1932, p. 64)

Early childhood imagoes retain their bottom-up all-or-nothing qualities, making them ego-ideal mythical figures: heroes, heroines, gods, or devils of destruction. Perhaps the distinction depends on the quality of love or hate attributed to the mother and father. From later in his life, Freud gives an example of what it means to have a leader who hates: Kaiser Wilhelm II's birth-injured left arm led to rejection rather than love from his mother. Freud surmised that he persisted in World War I destruction to get even with his mother and everyone else who rejected him.

In many of his later writings, including the Wilson book, Freud made inferences about the ancestral morality of leaders. He blames over-strict superegos directed against children by ancestors transmitting their superegos rather than their ego maturity: "… a child's superego is … constructed on the model … of its parents' super-ego … it becomes the vehicle of tradition … the tradition of the race and of the people, lives on in the ideologies of the super-ego" (Freud, 1932, p. 67). Freud criticizes the Marxist view of history, which places the whole emphasis of raw aggression and cultural regression on economic deprivation, which leaves out generations of superegos that interfere with maturational experience.

For example, he says that leading up to and during World War I, the British, French, and Germans, as well as the Americans, all believed they had god's backing for economic narcissism, that is, colonialism. Leaders of each nation rationalized that they were bringing the rest of the world up to their level. Given that the warring leaders were mired in these delusions of grandeur, Freud hoped Marxism was a better bet, but he had to conclude that the Bolsheviks showed the same narcissistic grandiosity, even if it meant mass murder of those who would not submit—shades of the present.

Given the disconnect between cultural rules and most individuals, Freud thought that the state of resistance in most analysands must originate from their sense that you cannot trust leaders, whose morality is self-serving and false just like their parents' and the world at large. You can only trust your kin and your religion as the least of possible evils.

WHAT DOES THE ID KNOW ABOUT ANYTHING?

Such thinking brought Freud to distinguish a bottom-up id separate from ego identity and superego morality. The id has neither a sense of morality nor a sense of time past; it only holds wishes from one's whole past, and a source of energy for every new fantasy that still seeks to find satisfaction despite being repressed. The lack of facts in the id reminded Freud of Wilson, who lived out a fantasy of being heir to the power of god. In *Lecture XXXI*, Freud presents Wilson-like id-states as mere fairy tales:

> ... hill-country, plains, and chains of lakes ... Germans, Magyars and Slovaks ... the Germans, who breed cattle, live in the hill country, the Magyars, who grow cereal and wine, live in the plains, and the Slovaks, who catch fish and plait reeds, live by the lake. ... Woodrow Wilson would be delighted by it ... [but actually] ...Germans, Magyars and Slovaks live interspersed ... in the hill-country there is agricultural land as well, cattle are bred in the plains too. ... (Freud, 1932, pp. 72–73)

In the progress of civilization in nations, many tribes and religions coexist without unbinding their aggression. The fact of coexistence instills progress. But in the above example which Freud attributes to Wilson's blindness to reality, Freud could see how the peace negotiations partitioned Central Europe so as to destroy the ways in which diverse peoples had managed to co-exist together. Segregating people by religion or ethnic background facilitated their cultural regression into archaic fantasies. Largely for the Allies' economic benefit, impoverished peoples were deprived of their real identity and geographically segregated, leaving their resentment to fester in their myths and laying the groundwork for another war. When people are removed from their homeland, as in today's world, they long for home, because, in a syndrome of restitution, their myths are based on fantasies of remaining a kinship group.

MORE ID

Longing for his past, Freud imagines the id blind to the external world, only attached to internal stimuli that demand satisfaction:

> ... fortunately for the id ... in its blind effort for the satisfaction of its instincts ... the ego must observe the external world. Must lay down an accurate picture of it in the memory-traces of its perceptions, and by its exercise of the function of "reality-testing" must put aside whatever ... is an addition derived from internal sources of excitation. The ego controls the approaches to motility under the

> id's orders; but between a need and an action it has interposed a postponement in the activity of thought, during which it makes use of the mnemic residues of experience ... what distinguishes the ego from the id ... is a tendency to synthesis in its mental processes. ... (Freud, 1932, pp. 75–76)

Here, Freud shows how thought delays gratification until salience appears in reality. Executive control requires a synthesis of the effects of both hemispheres in receiving external and internal information, using past experience before setting behavior into action. His thesis accords with brain imaging of sequential activations of intrinsic functional circuits.

Freud imagines id instincts attaching to ego identity. He uses his structural brain analogy to configure how identity forms: the hemispheres identify LB words with RB imagery to synthesize subjective with objective identity, joining LB need motivations with RB perception of the external world. Looking within inner space with his mind's eye of reflection, he generalizes a model for how the brain's mind makes identity: "... object-cathexes spring from the instinctual demands of the id. The ego has ... to take note of them. But by identifying itself with the object it recommends itself to the id in place of the object and seeks to divert the id's libido onto itself" (Freud, 1932, p. 77).

Thus, Freud accounts for identification: If I have your qualities I can love myself even more! A person wishes to find satisfaction by taking on attributes of others who have that satisfaction. Freud wishes for a structural model to unite vision with abstraction—like processes of fantasy and identification. The intention of psychoanalysis is:

> ... to strengthen the ego, to make it more independent of the super-ego, to widen its field of perception and enlarge its organization, so that it can appropriate fresh portions of the id. Where id was, there shall ego be. It is a work of culture. ... (Freud 1932, p. 80)

Chapter 14 explored how culture binds or releases individual aggression. The brain's innate superego treats the individual as a member of the tribe, religion, or nation, who is willing to die for primal kinship; because culture maintains a collective phantasy of its eternal life. The notion of culture as otherness that includes self-identity manifests Freud's conclusion that we inherit superego over generations. Chapter 15 uses a cultural case history showing that in cultural regression terrorism expresses collective syndromes.

Chapter 15

Syndromes of Restitution and Retribution

The Tsarnaev Case

Chapter 15 reassembles the information in *How Culture Runs the Brain* to show how much innate brain tendencies, individual fantasy, family phantasy, tribal mythology, global technology, immature narcissistic greed, and mature narcissistic hubris contribute to forming regressive cultural syndromes. Using Freud's conceptual framework about the development of individuals in stages corresponding to cultural development, let's consider a case history of a family and cultural regressive syndromes leading to terrorism. As we consider the destructive effect of collective syndromes, let's join Freud in pondering how to educate ourselves to regulate global culture in ways that provides everyone the opportunity to survive and feel meaningful purpose:

> In circumstances which are favorable to it, when the mental counter-forces which ordinarily inhibit it are out of action, it also manifests itself spontaneously and reveals man as a savage beast to whom consideration of his own kind is something alien. Anyone who calls to mind the atrocities committed during the racial migrations or the invasion of the Huns, or by people known as Mongols under Genghis Kahn and Tamerlane, or at the capture of Jerusalem by the pious Crusaders, or even, indeed, the horrors of the recent World War—anyone who calls these things to mind will have to bow humbly before the truth of this view. (Freud, 1930, p. 112)

On April 15, 2013, Tamerlan and Dzhokhar Tsarnaev committed the *Boston marathon bombing* like a religious duty. Michael Reynolds, an expert in Near Eastern Studies who testified at Dzhokhar's trial, provides a good understanding of the regression in human identity that leads to atrocity. It is easier to see perpetrators as monsters than examining how cultural syndromes affect their identity structure to promote terrorism. Reynolds's comments echo Freud's reminder that there is another way to see the brothers rather than as savage beasts who employ raw human aggression as religious zealots:

> It is … a common pattern of members of diasporas to form, and cling to, romanticized visions of their homeland and its traditions. The Tsarnaevs' selection of books, videos, and discussion groups all bore witness to the fact that they understood Chechen to be a central part of their lives. The sports they chose to excel in, boxing and wrestling are … revered in Chechnya … sources of national pride. When one examines the Tsarnaevs, past and present, one can see how the political, cultural and religious tensions of the North Caucasus indelibly stamped their lives, shaping their world view and self understanding and pointing them in the direction they took. (Reynolds, 2013)

This final chapter looks at the motives and the regressed fantasies of the Tsarnaev brothers as a case history of terrorism. It emerged within their family's cultural context, a syndrome interacting with American culture. We have been seeing how an individual's life and sense of identity, including Freud's, are tied to his or her psychocultural place in the world. Here, I'll trace how regression in levels of cultural fantasy within a family can lead to terrorism. The Tsarnaev brothers chose to follow contemporary preachers on the internet, throwbacks to archaic heroes who corresponded to the regressed psychic imagoes in their fantasy life.

There was no indication that Tamerlan and Dzhokhar met anyone who set up their operation, which corresponds to what some people call lone wolves. Faith joins reality if virtual contact is made with someone who promises the reality of a life in jihad. Experiencing cult figures on the internet can make fantasy identity seem as real as the religious phantasy/delusion of life after death. Thus archaic myths seem to grow real again—like the DNA of a dinosaur—if the person takes a risk based on advice and training of those who seem to know the whole process. Al-Qaeda pioneered internet techniques for making contact with those who were giving up hope for finding their identity.

Cultural disasters—war, genocide, and forced migration—cause trauma and loss that leads to individual PTSD as well as to collective cultural syndromes. Archaic phantasies become the source of identity change and revolutionary membership in times of cultural upheaval. Freud showed that stages of individual human development correspond to stages in the cultural development of group identity: nuclear families, tribes, religions, nation states, and globalism. Fantasy-based opposition to prevailing culture leads to character distortion, neurosis, psychosis, and collective syndromes.

But how much affiliation remains at each group level of identity development? Those allegiances are critical to the survival of individuals and their families. Migration mixes identities and produces confusion—a Babel of tongues. I don't think the Tsarnaev brothers, who each felt inhabited by more than one identity were insane, but their deployment of love and aggression in their identity fantasies must have felt bewildering. Subcultures become syndromatic when most of a population feels conflicting allegiances to family, tribe, religion, and state. If a person veers too completely from the prevailing cultural standards of normality, he or she feels estranged, with symptoms of poor self-esteem and defiance that may become antisocial. But what is immoral in one society is not in another. When nations impose hypocritical rules or oligarchic morality on diverse ethnicities, tribes, and religions, aggression may unbind.

CONDITIONS THAT FOSTER CULTURAL SYNDROMES: A CHECHEN EXAMPLE

In Freud's time, from nineteenth-century colonialism through twentieth-century globalism, nations were manipulated by rulers who displaced tribes, ethnicities, and religions from their homelands, depriving them of economic opportunity, imposing a festering submission and fostering revolutionary phantasies. The Chechens are an example of this. One-third of them were removed from their high Caucasus homeland in 1944, when Stalin loaded them into cattle cars and sent them to Siberia or Kyrgyzstan in the dead of winter with no resources. To undermine Islamic traditions and to change the ethnic mix, he replaced them with Serbs and Cossacks. Surviving Chechens developed a collective longing to return home, along with phantasies of retaliation. Further trauma occurred when Russia imposed republic status on Chechnya in 1993, and again under Putin beginning in 1999. Many Chechens developed a sociopathic morality rationalized by their wishes to return to what gradually became a mythic past. They followed what Freud described as ancestors' superego legacies that permeated their family myths.

Reynolds testified that Chechen survival often depended on criminal activity (Reynolds, 2013). For the Tsarnaev family, selling drugs, fixing cars without a mechanic's license, selling cars illegally, stealing, counterfeiting, and lying about assets to get public benefits were rationalized morally as ways the family could beat the imposed system. Reynolds ties the Tsarnaev's fascination with cars as instruments of survival and aggression to ancient ancestors who used horses to maraud on the Steppes. Among the myriad of public articles about the family that I used to piece together the puzzle of a family's diaspora, the most comprehensive was written by the Boston Globe (Jacobs, Filipov, and Wen, 2013).

The Tsarnaev brothers' parents, Anzor and his wife Zubeidat (an ethnic Avar from Dagestan) met as students in the republic of Kalmyk, which borders Dagestan. Tamerlan was born on October 21, 1986, the day after their marriage in the Chechen community of Kalmyk. They settled in Kyrgyzstan where two daughters were born: Bella in 1988 and Ailina in 1990. After the Soviet Union dissolved in 1991, Anzor moved his family to Chechnya, finally its own state; but he had to flee under political duress back to Kyrgyzstan, where Dzhokhar was born on July 22, 1993. Pointing his boys toward his heroes, Anzor named Dzhokhar after Dzhokhar Dudayev, the first president of Chechnya, later assassinated in 1996 (Reynolds, 2013). Tamerlan was named for Tamerlane, who used raw aggression to conquer the Asian Steppes in the 1500s.

Anzor was an impulsive, boastful, would be revolutionary, increasingly brain-damaged from bar fights. He was born in 1964 in Kyrgyzstan where his family had been deported by Stalin. His father collected scrap metal to sell, but was killed when a piece of metal he put in his old truck turned out to be an exploding shell. Of his seven siblings, three sisters and one brother, Ruslan, became lawyers. Anzor pretended to have law training, wielding a certificate of employment from the Kyrgyzstan prosecutor's office, which he used to impress people.

Anzor was traumatized in the mid-1990s when his attempt to trade in illegal tobacco ended by his being badly beaten by mobsters whose territory he invaded. He took his family to Zubeidat's Dagestan, where he worked as an unlicensed mechanic. In Kyrgyzstan, Chechens hid their Islamic beliefs, but in Dagestan, Avars openly practiced their Islamic culture. When Putin came to power in 1999, Anzor as a Chechen, worked against the Russian invasion of Dagestan. Ruslan Tsarni, his brother, later said Anzor was intimidated and tortured by the new Russian regime. Helped by Ruslan who was already moving freely through the golden door of America, Anzor, Zubeidat, and Dzhokhar came to the United States in April 2002 on a tourist visa. They were allowed to stay in the United States on the basis of Anzor's request for political asylum. Nine-year-old Dzhokhar learned idiomatic English and skipped a grade by the time Tamerlan, 15, and his two sisters, left behind in Ruslan's care for two years, joined them.

Most media sources view Ruslan as a family member who took a kindly uncle's position toward his brother's four children. Yet he had significant U.S. connections. His CV reveals he worked for USAID, Halliburton and other oil exploration, and international financial corporations in and around Kyrgyzstan. He established a *Congress of Chechen International Organizations,* which was registered at the address of Graham Fuller, who was CIA Chief for the Near East and South Asia in the 1980s; and in Afghanistan he formulated American policy that undermined the Russians. Under Reagan he engineered the Iran Contra Scandal. Ruslan was married to Fuller's daughter in the mid-1990s when both were students at Duke Law School, and they divorced in 1999 (Giambrone, 2013). Given Ruslan's connections, it is hard to evaluate his influence on Anzor and Tamerlan.

WRESTLING AND BOXING WITH AMERICA: ACCULTURATION OF THE TSARNAEV MEN

The pressures of the family's syndromatic past worked on their capacity to deal with their American life. The psychiatry resident, Dr. Alexander Niss, who treated Anzor for PTSD from 2003 to 2005, testified at Dzhokhar's trial that Anzor had auditory hallucinations, visual flashbacks, nightmares, and paranoid thoughts of KGB operatives following him. Niss treated him with antidepressants, antipsychotic medication, and sleeping pills; and also had some sessions with Zubeidat.

Soon after the family moved into Cambridge, Anzor started a car mechanics business in a friend's back lot, providing him and family members with cars and some money. But, as Anzor gradually lost hope of becoming an important Chechen person, he shifted his narcissism to Tamerlan, trying to make him a world champion boxer (Jacobs, Filipov, and Wen, 2013). He monitored Tamerlan's training runs through Cambridge streets, riding behind on a bike, goading him with advice. He even followed him into the Somerville Boxing Club, climbing into the ring to manage his sparring workouts.

Reynolds (2013) points out that Tamerlan's ring behavior was typical of Chechen men. Tamerlan would keep his arms by his side until the last minute—like Muhammad Ali at his prime, acting casual as if knocking out his opponent would be no big deal, since he himself was almost invulnerable. Although smaller than his son, that was how Anzor had been as a youth. Tamerlan was the New England Golden Gloves heavyweight division qualifier to contend for representing America in the Olympics, but he lost his first competitive match, before he could be disqualified for not being a citizen. Father and son felt the results were rigged.

FALLING APART, REGRESSING TOGETHER: THE WOMEN IN THE FAMILY

After a few years of living in Cambridge, the family began to fall apart as the two girls began acting out sexually and criminally. When Bella, the older sister, started seeing a non-Muslin Brazilian boy from her school, Tamerlan was told to intimidate him—which he did with a punch in the face. Bella was arrested for selling marijuana. Ailina used Ruslan's wife's name to register her car and get a false driver's license; then she drove her friends to a restaurant where they passed counterfeit bills while she stayed in the car.

Uncle Ruslan arranged a marriage for Ailina in 2007, moving her to Washington State where her husband lived and where Ruslan first lived when he came to America. Ailina called the police after her husband supposedly tried to strangle her; although she didn't press charges it was grounds for a quick divorce. Bella also had an arranged marriage in Kazakhstan that lasted just a year. While they were away it seemed the legal difficulties for these two young mothers melted into rationales of innocence.

Both daughters and their baby sons returned to the family home in Cambridge by late 2008; but, getting on Zubeidat's nerves they all moved away to New Jersey by 2010. Indeed, neighbors said Zubeidat yelled at her family members loud enough to be heard for a block. Now Zubeidat could focus her love on Tamerlan. He had met Katherine in a club, while living with a non-Muslim girlfriend whom he punched. Katherine was a 17-year-old from a well-off American family who married him in 2010, converting to Islam when their daughter was born. They lived in the Tsarnaev's cramped apartment. By then, Anzor had suffered severe head trauma requiring a month hospitalization while fighting with members of the Russian Benevolent Society, which damaged his capacity to work on cars.

TAMERLAN'S IDENTITY REGRESSION TAKES HOLD OF THE FAMILY

As Anzor waned, Tamerlan felt the urgency of his father's wish for him to maintain the unconquerable aggression of the indomitable Chechen hero—the Tamerlane of the Asian steppes whose aggression was invincible. For Zubeidat he was to show the Avar purity of learning and fundamental love of the true Islamic god. She was ambivalent about jihad. As Tamerlan increasingly lost control of his raw aggression in 2011, she got him to pray together with her. Yet, she also insisted that the Americans were behind the 9/11 attacks in order to justify killing Muslims. As Anzor declined, she provided a *folie a deux* with Tamerlan based on increasing symbiosis and fundamentalism. Ruslan later blamed her for changing Tamerlan from a more modern Muslim to a jihadist.

As the family's regression increased, Zubeidat told friends that Tamerlan seemed to have two people living in him—as she did herself. Recovering her Avar ethnicity and her Dagestan persona, she pushed Tamerlan to abide by Salafi law, while she traded in her modern clothes for modest Islamic hijabs. Stopping her paid cosmetic work in her home, she studied Salafi with Tamerlan, who gave up alcohol, marijuana, and fancy clothes to follow the works of Sheikh Abu Malik Abdulhamid al-Juhani who preached on the internet that true Islamists should obey the fundamental Quran (Reynolds, 2013).

FULL IDENTITY REGRESSION SETS IN

Tamerlan's friend, Brendan Mess, and two Jewish gym buddies were murdered on September 11, 2011, the tenth anniversary of 9/11 attacks, their throats slit almost to the point of head dismemberment. Scattered over their remains was nine pounds of marijuana and five thousand dollars. Tamerlan, who dealt marijuana, was suspected of being in on the killing. Another gym friend of his who suddenly moved to Florida allegedly signed a confession of taking part in the killing with Tamerlan, but he was said to have attacked the FBI agent he was working with: so the agent *had* to kill him in self-defense. At the time, Tamerlan was said to be investigated by Russian authorities for possible terrorist activities, which held up his application to become an American citizen. Yet he was not on any American watch list.

Later in September 2011, Anzor and Zubeidat went before a Massachusetts judge to end their 25-year-marriage. Anzor said he and Zubeidat had already been separated for six months—but he may have been in Dagestan setting up a new living situation. Soon after divorce, he and Tamerlan went to Dagestan where Zubeidat later joined them after she was arrested for stealing $1600 of cloths from Lord and Taylor. Back in Dagestan, where Islamic resistance to Russian authority and jihadist sentiment was high, Anzor had enough money to buy an apartment and set up a business. Tamerlan, acting like a man about town gave money to many beggars.

Returning to Cambridge in 2012, where Katherine (now Karima Tsarnaev) still worked to help support him and their child, fully bearded Tamerlan was immersed in his fundamentalist jihad identity, blaming the United States for incursion into Muslim affairs in Iraq and Afghanistan. He was often speaking by phone with Russians. Yet he told a friend he was being pushed by *someone* in his brain to do things he did not want to do. His identity was split. Without his parents, Tamerlan lost welfare and section 8 housing, the landlady raised the rent, and Zubeidat stopped sending rent money from Dagestan. America was catching up with the brothers.

JAHAR REVERTS TO FAMILY MYTHOLOGY

Jahar was the only member of his family who had developed a comfortable American identity. Peter Payak, his high school wrestling coach, had a strong

Americanizing influence on Dzhokhar. Peter had gotten the other wrestlers to call Dzhokhar, "Jahar" (Reitman, Janet, 2013). Jahar was the popular captain of the wrestling team for the last two years of high school. An American father figure to Jahar, Payak's devotion to the boy was matched by Jahar's devotion to his team mates and friends. Jahar would tell them he always had their back. He could be called upon in the middle of the night by whoever needed him—even for marijuana. Jahar was open in tweets with American, Chechen, Russian, and Islamic friends. But American friends never came to visit him when Tamerlan was home.

Payak loved running in the Boston Marathon, telling Jahar his times for each of his 12 runs. In February 2013, Jahar visited the wrestling team, playfully exerting his strength and skill with team members. It was a cruel goodbye to his American identity, for he had learned that Peter would be waiting for his own son to cross the finish line. But he went ahead and planted the bomb. Ironically, the explosions were to render Peter almost deaf in one ear. Even so, after his son told him the picture of the fellow with the back pack in the media pictures placing the bomb was Jahar; Peter let the FBI use his name to reassure Jahar that if he came out from under the boat tarp, where he was hiding after Tamerlan was killed, he would not be shot. Peter said Jahar was "seamless as a billiard ball," a good kid—if a monster.

By 2013 Dzhokhar felt marooned at the University of Massachusetts at Dartmouth, where he was failing courses and running up $20,000 debt for each school year—although his thriving sales in marijuana to students provided plenty of spending money. In December 2012, he became an American citizen, but he lamented in his tweets that he missed Dagestan and Chechnya: mother and father. A few days later he tweeted that 9/11 was an inside job, engineered to give the United States a rationale for killing Muslims. That was the litany of his parents, Tamerlan, and many internet preachers. Michael Reynolds noted that Anwar al-Awlaki was Dzhokhar's favorite influence. He had a reputation as trustworthy, yet among Mujahedeen he was thought a master at converting disciples (Reynolds, 2013).

Dzhokhar stayed with Tamerlan in January for weeks after the brief 2012 school Christmas holiday, tweeting that he had three zombie apocalypse dreams, and a dream about the end of the world. In February 2013, he tweeted a dream/fantasy that he had killed Abraham Lincoln. It occurred during a daytime "nap" of two hours in which he had a strange sense of intense reality. These were symptoms of identity splitting: awareness that his inner world and his American identity as he knew it was coming to an end. All the while, Tamerlan was preparing jihad. In February, following the Al-Qaeda *recipe*, "make a bomb in the kitchen of your mom," and reading about jihad military tactics, he bought large amounts of fireworks explosives, weapons, and ammunition; he took Dzhokhar to a firing range in March. Like Hamlet, Dzhokhar tweeted that decisive men have to take action; you can cheat in your studies but you can't cheat about going to god's heaven. Loyal to Tamerlan he was acting out family myths of Chechen and Islamic heroism, while his American refuge was melting away in the April spring.

"MAKE A BOMB IN THE KITCHEN OF YOUR MOM"

On April 15, 2013 each brother placed a homemade pressure cooker bomb, hidden in a backpack, on the ground close to the finish line of the Boston Marathon. They followed an Al-Qaeda on-line recipe for bomb-making by filling two pressure cookers with nails, pellets, and explosive fireworks powder; they left them at ground level a block apart, using toy-car electronics to set the off bombs 13 seconds apart. They lingered to watch the results: two adults and a child were killed, and 264 adults and children were wounded, many losing their legs.

Two days later, on April 17, planning to get a potent weapon for their next venture—perhaps Times Square—they drove to MIT in Dzhokhar's green Honda, where Tamerlan, now in a frenzy of aggression, shot and killed a campus officer, while trying to steal his gun. Fleeing on April 18, a desperate Tamerlan carjacked an SUV, bragging to the owner he was the Boston bomber, took him hostage, and forced him to get the maximum $800 from a cash machine. With Dzhokhar following in the Honda, Tamerlan stopped to get gas; his hostage escaped and called the police, which led to a shoot-out. Riddled with bullets and out of ammunition, Tamerlan got out of the SUV; Dzhokhar, aiming his car at the police to save Tamerlan, hit and dragged Tamerlan's dying body in the process.

Dzhokhar escaped, hiding overnight under a tarp covering a boat in a back yard. The owner found him bleeding and called the police, who shot him repeatedly under the tarp. Fortunately, Dzhokhar emerged alive—fortunate because he lived to describe what he and Tamerlan had done, how they did it, and why. They were not part of a terrorist organization. They had carried out Tamerlan's jihadist decision to kill Americans in reprisal for the collateral killings of Muslims in Iraq and Afghanistan. While in hiding, Dzhokhar scratched out two defiant messages on the inside of the boat. First: since the U.S. government kills innocent civilians, he had to retaliate: killing one Muslim you kill them all—they are of one body. Second: "Fuck America": American idiom renouncing American identity. Dzhokhar's Muslim and Chechen family identity were reunited.

SYNDROME FORMATION

How did the Tsarnaev family foster an act of dehumanized anti-American terrorism? The brothers expressed emotionally disturbed identity splitting in their American experience. Feeling more than one self contending for control of mind and body creates a tumult of fantasies that seek correspondence in reality. Anxiety triggering identity repression can be alleviated by drugs. Perhaps marijuana can present the illusion that each fantasy, with its own attendant past context of identity, can synthesize an image of identity satisfaction. This seemed to be the case in all four Tsarnaev children. Anzor and Zubeidat were quite different in their organizing fantasies and in their superego ideals. Anzor was pragmatic and opportunistic from day to day, yet he needed some narcissistic triumph to justify his Chechen longing for a permanent home. Zubeidat tried to be a super-attractive modern woman, but her Avar self wished to be austere and perfect. To compensate for regressed split-identity, a person searches for a narcissistic mirroring of one singular cultural self.

As context for cultural syndromes, consider that each level of group membership, from nuclear families, to extended families, tribes, ethnicities, religions, and nations inhabits a global world that sets the economic circumstances for most of its inhabitants. Each person's individual identity is compounded from their identifications within their array of memberships. It is distressing to everyone to feel pressure to absorb conflicting identity. I define a cultural syndrome as a subculture's collective aversion to an alien morality impacting their survival.

TRIBALISM VERSUS GLOBALISM

The source of love and familiar identity remains in symbiotic kinship, losing primal intensity as consciousness progresses up the developmental scale of the cultural hierarchy to what has been called the global village. From the top-down that hierarchy corresponds to the DMPFC's developmental organization of reflection on identity. Leading nations set the moral organization of the global economy. Narcissistic leaders controlling most of the world's resources have a disproportionate role in formulating global trickle down to subcultures that comprise local communities. When survival is hard, local leaders try to formulate a familiar morality directed toward maintaining the survival of their community's families, tribes, and religions.

In Diasporas, wars, genocides, and refugee camps, immediate survivors are often the most sociopathic; those who feel little or no remorse in others' death. That leaves collective PTSD to other survivors, whose children enter syndromes of loss, revenge, and compensation. Phantasies merge with myths that animate the revolutionary sociopathy of the down and outcast. There is a fixation on symbiotic love and hate in some families of revolutionists as members search for ways to bring collective phantasies to fruition.

Sociopaths and revolutionaries join forces. The Tsarnaev children identified with imagos of their forbears. Dzhokhar seemed the most loyal, showing the most love for his kin. Hiding inside the boat under a tarp, Dzhokhar wrote poetically: “Know you are fighting men who look into the barrel of your gun and see Heaven .... We are promised victory and we will surely get it” (Valencia, Milton J., and Wen, Patricia, Boston Globe, March 10, 2015).

Killing those who belong to a tribe that kill your kin may feel justified. From that perspective, many Al-Qaeda and Isis jihadist leaders including Osama bin Laden wrote poetry. *Battle Lines* in the New Yorker described poetry of jihadists, strangers who mean to reclaim the world from nation states that cast them out (Creswell and Haykel, 2015/06/08). One jihadist phantasy is an Isis without boundaries. Creswell and Haykel write that outsiders become insiders in poems of jihad inspiration. Islamic monorhyme poetry—simple lines with single rhymes—make a repetitious beat of aggression that motivates jihad—in a cultural redemption from humiliation. They report on bin Laden writing poetry as if in his young son’s view: The eternal wanderer alienated from tribe gives his moral strength as a legacy of jihad. Terrorists may use poetry to bring their children into their own syndrome. In one bin Laden poem, his son’s defensive response to the aggression against his people is like surviving a rain of U.S. missiles, hurling each one back in jihadist transformation. It is an ode for the dead warrior, revenge for the death of their identity, and symbiosis with the homeless. Those are themes that Dzhokhar adopted in cultural identity regression to his family’s syndromatic, jihadist battle lines.

In Isis there is an air of the archaic Bedouins who once were nomads without a fixed home, also of Genghis Kahn and Tamerlane, traversing the Asian Steppes, killing all who failed to recognize them as authentic leaders entitled to live and multiply. Isis seems to glory in taking on the identity of the stranger who becomes a follower of Muhammad. The outsider lives out the romance of becoming the insider, the progenitor of a new order. That is the nature of the internet invitation to join the religious ranks.

CRIMINAL DESTRUCTION AND RECONSTRUCTION OF THE WORLD

I think civilization does depend, as Freud thought, on the maturity of leaders who work together for the sake of everyone’s survival and quality of life. But as Freud also realized, overly narcissistic leaders treat the world as if it belongs to them. That breeds a criminal mentality that can be rationalized as a sucker is born every minute. A leader has to have empathy for everyone;

otherwise the system for mutual survival breaks down. When criminal oligarchs lead the global economy, market needs are suppressed and economic policy is dehumanized. It is clear enough that this is the case in many parts of the world. Using rationalized lies, cultural regression passes back through the stages of childhood.

When honest logic fails, cultural regression works against civilized life. With significant help from President Obama, the United States still had a functioning democracy that can work within a more or less fair playing field. But political opportunism can overwhelm cultural stability. Binding aggression requires honest dialogue and compromise. When a culture gives way to sociopathic narcissism, aggression is unleashed in chaotic ways. If the poor bear the weight of bias against their needs, they may revolt. From an archaic revolutionary perspective on oligarchs, terrorism is a dose of their own medicine.

The Tsarnaev's belief that 9/11 was an inside job expressed an illogical rationale of a cultural dynamic unconscious posing a metaphor of their threatened survival. Projective defenses explain away the degraded condition of estranged groups that feel futureless. Revolutionary leaders who are capable of humanitarian rationales may claim that like Tamerlane, Genghis Kahn, and Christian crusaders, they fight for god's morality—even sacrificing innocent people.

When a culture is so regressed as to lack coherent leadership, honest policing, and economic support for the poor, it invites sociopaths and visionaries to sell fantasies and drugs that make those fantasies seem real, and to use murder and intimidation to prove that they are in charge. If the country has remnants of maturity, decency, and economic coherence, the clever ones may form a Mafia kind of morality that binds aggression with loyalty to the leader. Looking at Mexico and Columbia, for instance, it seems to me that when drug leaders reenter society with money and influence, they may try to rebuild institutions. Their nation may benefit by becoming more involved in the global economy.

In Freud's view, collective cultural maturation happens when a majority of ethnic, tribal, and religious groups living together are willing to bind their aggression so that their own kin will survive. In renewed nation building, they are led to compromise by leaders who persuade them that giving up a portion of identity to a larger amalgam of culture produces a better life for all. Strangers can be experienced as *just like you and me*.

Perhaps revolutionary political cycles correspond roughly to Plato's Republic: An *aristocratic* philosopher king, for example an idealized father leader, uses reflective wisdom to rule wisely; he is followed by a generation of lesser men or sons who rule by logic, but, giving in to immature narcissism, care more about their own advantage than other's rights; then comes *oligarchy* as wealth is concentrated in the hands of fewer men who gorge to their own satisfaction, while neglecting everyone's safety; then *democracy* erupts, but the people lack the capacity to guide themselves; and finally comes *tyranny,* with a leader so paranoid that he loses the capacity for love and becomes obsessed with killing whoever threatens his hegemony—and the cycle repeats.

THE RULE OF THE NOMADS

Freud insisted that civilization results from the sublimated love of men working together for progress and binding their aggression for one another. Binding aggression also requires the good will of women, who do not feel they are owned by men, as some fundamentalist religions hold. The historical evolution of culture also included Asian Steppe nomads who killed those who would not submit to being governed. In this way, Genghis Khan, the scourge of the great grassy Asian Steppe highway between West Europe and East Asia, gradually cleared the Silk Road for safer trade. We have seen that the Buddhists facilitated bound aggression between settled and nomadic cultures. Their relaxation techniques and philosophy soothed defensive aggression and allowed the benefits of trade and mutual learning to enhance life on both sides.

Tamerlane, the last great warrior of the Steppes, was a narcissistic aggressor who conquered nomadic and settled cultures alike, from China and India all the way to the Mediterranean, always to his economic advantage. He saw himself as the master of trade. The Silk Road became his road. Tamerlane was said to be a descendant of Genghis Khan on his mother's side, but of a sedentary tribal leader on his father's side. Although his father converted to Sunni Islam, Tamerlane killed Islamic peoples with the same ease he brought to massacring the population of any of the large cities that stood in his way. He was the tyrant of trade, while admitting all manner of strangers to his conquering forces. He was Tamerlan's namesake.

From a Freudian perspective, cultural love and aggression are social projections of the raw, biological sources of instinctual life. They organize individual experience throughout stages of individual development. Experiencing communication from one another, we peer into sources of our own personal identity. That is a lot of primal energy to exchange. When there is too much unbound aggression in too many people and in too many cultural institutions, those who sublimate their love and bind their aggression find navigating culture to be a huge and futile effort. When most people need guns, culture itself has become the enemy.

UNFAMILIARITY BREEDS CONTEMPT

Everywhere we look, whether to brain imaging, life experience, or group status, we find that cultural familiarity breeds content; while unfamiliarity breeds contempt. We feel bias within 0.2 seconds when we meet a stranger, before we become aware of it at 0.5 seconds. The more complex collective life becomes with larger aggregates of strangers, the more we all have to bind our biased antipathies in order to find common ground. That is where cognition enters the picture—it alters defensive responses by binding aggressive affects. Our leaders have to convince us that maintaining a viable culture takes work by everyone for mutual survival. Whereas criminal oligarchs jeopardize collective survival, mature leaders are aware that global compromise, through trade, diplomacy, and mutual security, is necessary for tribes and religions to survive in their subcultures.

Large-scale contact with strangers usually meant either threats to survival or good fortune in sharing and trading. We have seen that the cultural evolution of the human brain reflects that ambivalence. When people migrate they are vulnerable to an excess of ambivalence, particularly since they often move when survival is too tough. Statistically, we can estimate that one in seven people in the world are on the move, most within their country, but 3–4 percent to another country. In 300 BCE there were only 250 million people in the world; yet migrations brought Rome to an end. Now there are well over 7 billion people, with 3–4 percent emigrating, living in refugee camps, or living as virtual slaves no longer needed in their host country. Now, 250 million people have no place to live. For many of them, their identity resides in phantasies.

PUBLIC AMBIVALENCE

Leaving Jahar and American idiom behind, Dzhokhar apologized for the irreparable damage he did to so many people. In his posttrial statement he said he prayed to Allah for their relief from the suffering he caused them. He said he tried to feel their broken lives as he listened to them and their family members testify at his trial, by immersing himself into Islamic awareness of their lives. Dzhokhar turned to his Islamic identity, not as a jihadist, but as a person who feels love through Allah.

Does Dzhokhar deserve the death penalty or the life penalty? He was in a state of detached trauma during the whole episode, fascinated by Tamerlan as the source of family belief. A hypnotic state is the best description of a person like Patricia Hearst holding her gun for the Symbionese Liberation Army. Dzhokhar too was in his brother's thrall, just like an archaic youth following the leader's direction. Terrorists often recruit children to make them killers. Children are ready to adopt a new culture with its own morality. Little remorse can be engendered while joining in to the identity of new family romance heroes.

Isis leaders, for instance, can give suicide bombing missions to terrorists whose culturally syndromatic families have regressed to the point of belief in jihad symbiosis. Giving up their life to kill the enemy is rewarded by eternal love in paradise. That is a reason to use brothers or sisters who share the same cultural beliefs as suicide bombers. This concept also explains why people will vote for those who are inimical to their interests; it is common practice to lead them to believe they have the same enemy.

In a family romance pursuit of individuated identity through their accumulated day-dreams, latency children imagine narcissistic heroism and empathic altruism. Their cultural superego may feature ideal communities, or with less cultural binding pursue a world like the *The Lord of the Flies*. In cultural regression to unbound narcissism, Isis leaders can live out their phantasies of becoming leaders of new culture. They wish to form their own global hierarchy, or at least their own nation. Their aim is to join the global culture by exploiting, owning, and imaginatively creating a revolution that includes those who long for archaic cultures. Individual regression to archaic stages of fantasy joins phantasies of how things were and how they should be.

Alan Dershowitz holds that imposing the death penalty against Dzhokhar would make him a martyr; those seeking paradise would find him a role model. Politically, and in the popular mind, that poses the question of whether his terrorist act is judged by wartime or peacetime criteria. Since Dzhokhar is an American citizen, that also begs the question of how codes of law protect culture from the unbinding aggression that injures the culture itself.

CONCLUSIONS

Each of us synthesizes an identity over and over, from fantasies that are residues of our developmental stages, especially childhood stages. Freud's top-down ladder shows how we view each change in retrospect, so that we build our own structure of being. To Freud's perspective on individuated identity, we can add neuroscience findings that provide brain imaging portals into the construction of identity. *Cultural* identity is experienced as a virtual reality: Each person enters and internalizes culture through the auspices of some leader who represents social reality, every culture has forerunners who constructed laws from cognitions, judgments, and rules of behavior and those qualities bind the myths and wishes still percolating in our cultural unconscious. Emotion generalizes the archaic laws of the wild, the mammalian forms of affective communication that show how to survive as a group; while fantasy is all about imitating and identifying with others who show a way to satisfaction. According to Freud, narcissism brings the aims of instinctual pleasure vested in others back onto the biological self, like Narcissus loving his body image reflected in the water—or her body image reflected in culture's mirror.

Death as the ultimate reality means that we build our personal castles in the air. Living in a culture with everyone else in that dilemma makes us vulnerable to whatever our culture offers us—it is our milieu for better or worse, and when it becomes worse for everyone, we hunker down and build cultural syndromes, group-centered myths to calm our anxiety, as long as possible. From tribal times to the present, being a member of the family group conveys the symbiotic benefit of access to wishful pleasure. For access to that relief, a young child must learn to imitate older kin in speech, gesture, affect, and use of aggression. In a bound state you give or receive kindness; in the unbound state you experience bullying or being bullied, owning or being owned—you are master or slave, sadist or masochist, potlatch-giver or thief.

Through the ages of human evolution, in the brain's distinctions among cognizing reality, feeling the outer and inner body, and relating through emotional communication, leaders and shamans have resorted to unbound aggression to fulfill their narcissistic identity wishes. The immature narcissism of sociopathy may become a refuge, even a collective syndromatic milieu. Freud found that youngsters who feel unprotected, may develop neurotic solutions to unbearable experience—just like their parents.

In the early days of culture, some shamans, as mature narcissists, observed that owning holy land and controlling survival resources was the way of the animal gods who led tribal and religious culture. So they communicated cultural stories as living myths, like young children who observe others using what seems like magic utterance to control what happens. Freud called that obsessive magic. In similar fashion, reciting Catholic liturgy, or memorizing the Quran or Talmud, conveys magic cultural power. Evolving religions had shamans who cared for the emotional needs of cult members, and some were elevated in rank to actually regulate holy lands and other survival resources, while sharing the benefits of their narcissistic leaders—like a trickle-down blessing. Grandiose imitation leading to widespread cultural destruction is the path of immature narcissists, whether religious or cult leaders, or political or economic oligarchs.

As Freud saw it, the fantasy of a personal incorporation of god as a totem, or a phantasy of group cohesion identified with gods and goddesses as sources of symbiosis, may lead to religious delusions. Too many leaders go to war to defend the proposition that god gives them the right to be better and richer than others. That was Freud's story of World War I, and could be our contemporary story. He mistrusted U.S. economic hegemony, especially Woodrow Wilson's claim that god had given the United States the Christian strength to prevail. For him, Wilson embodied the grandiosity of identifying with Christ.

HOW IS A CORPORATION A PERSON?

Exploring the sources of cultural syndromes, I have largely ignored other disciplines that examine the conflicts that roil civilization. Although unqualified to write about other disciplines, I'll take a few stabs about the effect of constitutional history and corporate influence on U.S. national and state culture. The founding fathers balanced contending economic interests, contending colonial powers, contending religions, and contending populations, among them sects, Indian tribes and slaves to establish a collective identity. Compromises in the formation of the Constitution and its Amendments were written in the language of idealized ambiguity, in order to provide the illusion of unity. The United States, that is, the inhabitants of those states, had to accept one another in order to maintain a collective emotional allegiance strong enough to bind selfish raw aggression. The Constitution's preamble is the mythic American Dream:

> We the people of the United States in Order to form a more perfect Union, establish Justice, insure domestic Tranquility, provide for the common defense, promote the general Welfare, and secure the Blessings of Liberty to ourselves and our Posterity, do ordain and establish this Constitution for the United States of America.

The First Amendment protected the legal rights of free speech, free press, free religion, free assembly, and petition for redress; and it protected the collective national identity against revolutions that could be fostered through individual speech. Thus, advocating change by force is a punishable crime of sedition. If the First Amendment bears on citizens' rights to life, liberty, and the pursuit of happiness, the Second bears on the right to use firearms for protection. Interestingly, the first two amendments accord with a Freudian view of the instincts of love and aggression; and how we bind those instincts developmentally to show concern for others. According to Freud, property marks the avidness of early childhood, and religious belief signals a slightly later stage of identifying with older kin and with a sense of god in pursuing one's own wishes. Perhaps god is the ultimate ambiguity in the brain's prefrontal capacity to generalize.

A nation's mythology screens the aggression unbound in the revolutions that created it. Freud believed that in the evolution of a culture, return of the cultural repression of aggression against subpopulations haunts its future. For the United States that aggression included building economic hegemony in Southern states on the backs of black slaves treated as commodities—not to mention the genocide of Indians. Biased treatment of the black population continues to be a factor of hidden guilt and shame inherent in collective American identity.

Developing American identity after the Civil War led to three amendments to the Constitution that affected Reconstruction in the South. The Thirteenth (1865) abolished slavery, the Fourteenth (1868) gave equal protection under the law to all citizens, and the Fifteenth (1870) prohibited voting discrimination against citizens—excluding women. These amendments reconfigured American identity toward the ideal of making all citizens equal under the law, but their enactment was often blocked by states' rules and separate legislation putting obstacles in the way of blacks' full representation. In 1925, fear of the Bolsheviks making inroads on U.S. security led to the Supreme Court *incorporation doctrine*, an application of the Fourteenth amendment and Bill of Rights to states as well as federal law. Down to today, equality among individuals regarding their national rights remains eroded by state interpretations of the law.

Immature narcissists have long used corporate influence to support their own interests and biases. The prime example is a U.S. Supreme Court Decision in 2010, based on the Fourteenth Amendment, holding that a nonprofit corporation is legally a person; a decision that was extended to for-profit corporations as well. Corporations are now treated as individuals in terms of equal protection under the law, so they can use unlimited funds to attempt to influence elections. The 2010 ruling has been amplified by the suit brought by Citizens United (a super PAC largely funded by the Koch brothers), which produces films, books, and ads meant to influence political decisions. Citizens United produced media to vilify the Democratic Party part of the Republican attempt to defeat Hillary Clinton in her first bid for the presidency. When the effort was blocked by the McCain-Feingold law against corporations attempting to influence elections with media, Citizens United sued the FED (Federal Election Commission), winning their case in the U.S. Supreme Court.

The business of using media, lobbyists, and politicians to influence the public with false cultural information is an overt form of sociopathy used by immature narcissists. Since 2010, Citizens United and the Koch brothers have made huge inroads in state elections, using money and influence to support conservative Republicans. Indeed, CU chairman David Bossie became the nominal manager of the 2016 Trump campaign. The fabulously rich Koch brothers provide an explicit example of immature narcissists, for whom the profit motive is the only motive, so that environmental conditions that affect everyone's long-term future and the future of future generations are discounted. In comparison, mature narcissistic leaders contribute to forming a global economy with a more enlightened trickle-down effect, as well as authentic concern for the world's poor. They have no problem in accepting blacks, Latinos, or Asians in their cohort. Global narcissistic individuation, most iconic in the internet, has contributed to a world view incompatible with the adaptation of tribes, economic subcultures, and religions.

In terms of individual and cultural identity, incorporation has another meaning; it is a Freudian term for infant identification with others in the formation of personal identity. The O.E.D. defines corporation as a body of united persons, legally organized to act as a single individual, like an artificial person, a body with rights to perpetual succession: it thus signifies a cultural phantasy of group symbiosis. Such a delusion translates into belief in a god-given right to live in a culture forever, a phantasy that animates Isis or Al-Qaeda as much as any corporation.

Corporate capitalism is a top-down syndrome of rationalized grandiosity, reciprocated by bottom-up cultural syndromes of restitution and revenge. Cultural institutions are projected brain functions and structures that develop an organizational dynamic of their own, interacting with individual brains, individual psychology, and social communication. I cannot agree with the belief that any functioning brain is the sum of its physical components. Culture runs and ruins the brain, progressively toward globalism, regressively toward tribalism. Given the vicissitudes of physical reality affecting all these interacting levels, evolution determines what sticks.

Epilogue

With soon to be 10 billion people competing for survival resources in a world that needs multinational cooperation, cultural identity effects require widespread understanding—or our nature will take its course. In the cultural regression of the last decade there is little belief in Freudian developmental dynamics. Instead, when collective biases liberate raw aggression, social conflict, and existential fear, many people regress into early childhood identity modes: tribal loyalty and distrust of strangers. Multiethnic fairness is replaced by a dehumanized denial of others humanity or by narcissistic entitlement of owning others.

The *great recession* of 2008, caused by large U.S. and European banks selling derivatives that inflated real-estate value combined with decreased income in oil rich nations due to widespread fracking, altered the balance of power among nations. Arab Spring revolutions failed, American hegemony was upended, and regression among unstable nations led to wars, genocides, and national disintegrations stoked by Al-Qaeda, Isis, and Taliban terrorism often fed by regressive forms of Jihad. Xenophobic fear ignited by massive migrations and terrorism led many nations, including the United States, to buttress their own identity. The political right in many nations launched populist aggression against migrant refugees.

In the prolonged aftermath of war, genocide, forced migration, and economic loss, a culture's shared identity often becomes syndromatic, leading to sadistic, masochistic, sociopathic, and revolutionary behaviors. Belief in religious miracles replaces a shared cultural understanding of why everyday life is disintegrating. As we saw within the Boston bombers' family, one source of enacting Jihadist terrorism erupts when phantasies of restitution and revenge idealize and imitate past heroes who had once established their culture's mythic survival.

The 2016 American election highlighted the effect of U.S. cultural regression when Donald Trump won the Electoral College votes to become president. His mission to restore American greatness resonated with white populists who had lost their jobs to cheaper labor in the global economy. Trump gave a voice to people feeling betrayed by the national government. Right-wing American economic, political, media, intelligence, and military leaders—some of them resentful, immature narcissists—were finally able to overcome the mature narcissists who espoused multiculturalism. Elite global-minded leaders who influenced policies for progressive cultural diplomacy lost their advantage, undone by having taken the lion's share of economic resources for themselves. By defeating the progressive elite, Trump reversed the humiliation that many overlooked groups had felt.

Making Steve Bannon his guiding policy light, Trump enhanced his capacity to falsify information, and to use the media to reinforce his winning strategy. Even after his election, that strategy led him to attack *The New York Times* as rigging his presidency against him—by telling grown-up truths. As we have seen, Freud understood that distorting facts by going to war with god on your side was a way to act out a death drive by turning morality against itself. Religious views are at the fulcrum of identity clashes in the interaction between logic of policy for cultural and global progress versus a more fundamentalist tribal morality. Fundamentalist politics are closer to tribal organization: Outsiders are bad, insiders good, which denies multiethnic groups a path to surviving through compromise by understanding their place in the larger world.

Like young children projecting their aggression, conservative Republicans during the Obama presidency projected their gloating schadenfreude, taking pleasure in blocking Obama's policies. Trump used their foibles to humiliate those who opposed him in the primaries. I think many who faded under his attacks could not answer him because of their Grover Norquist pledge not to vote for any budget deficit—a condition for receiving money for their own reelection. Their allegiance was not to the country but to defeating Obama and the Democrats. The Democrats did not effectively fight back, leaving Obama to manage on his own. Hillary's Democrats had hubris, feeling their ideals were superior.

Like Louis XIV's "L'ètat C'est moi" Trump believes his identity is sovereign. To the extent that he surrounded himself with a cohort espousing the same narcissism of total entitlement, he acts out a right to humiliate those who have been defeated as mere losers. He uses those whose loyalty he can count on—especially family members—like a Mafia structure for his leadership. For example, Jared Kushner, the son-in law married to Ivanka Trump, is like a chameleon of influence for Trump, changing color according to the environment. His father, a holocaust survivor, became a real-estate magnate convicted in 2005 by New Jersey Attorney General Chris Christie for illegal campaign contributions used in tax evasion and witness tampering. Jared is heir to his father's billions.

He treats Trump as the besieged father, which assuages Trump's narcissism. As a self-object to Trump, Kushner influences his decision making about whom to trust—and he advises him about how best to get even with those who harmed him. In a book written with William Bullitt about President Wilson, Freud analyzed the effect of a narcissistic president who left Europe in tatters after WWI. He and Bullitt showed a narcissistic leader's vulnerability to being influenced. Wilson had his Colonel House who would introduce small changes in Wilson's narcissistic policies, which would add up to House's influence; just as Kushner's influence on Trump adds up to Kushner's power.

Trump treats the nation as his own enterprise, for example, getting someone else to pay for debt incurred during real-estate trade. But leading the nation into high debt that need not be repaid (as he did in his casino projects) has different results. Weakening the dollar—the global standard of currency—can result in global economic depression. Similarly, pretending that environmental regulations are mere impediments to business uses up irreplaceable resources, which weakens the global cultural future of everyone's survival. His deregulation presages a potential, massive, Ponzi scheme.

His Republican agenda of cutting back financial regulations and going into debt to expand the economy also shifts more benefits to Republican-controlled state legislatures. Highly controlled state legislatures are more tribal-seeming, clannish, and populist and less inclined to global trade unless they can make their own state deals with foreign governments. Unlike California or New York, coastal states with more global trade, southern states take more federal benefits than they give up in taxes. Second amendment rights interpreted to include the use of war weapons to maintain autonomy are held as protection against too much federal regulation.

Populism in the south and the rust belt manifests innate tribal tendencies, quite different from California's Silicon valley technology which is oriented globally. When families survived on manufacturing jobs for generations, living in communities ruled by local corporate leaders, they developed a tribal attachment to their life. They had more clout and economic benefits than did American Indians or black people who had less legal protection as citizens. Globalism as a progressive form of multiculturalism always clashes with tribalism, just as maturation always clashes with early childhood identity.

A regressive culture based on bias and early childhood identity loses collective reflection, which leads to detachment from causal reality. Freud said that cultural progress depends on using science and technology—so does winning wars. The elite reflect on the physical and social sciences to impact culture. Without reliable systems based on science, people have no way to deal with problems that sooner or later cause a lack of necessary survival resources. For instance, statistics that measure the state of an economy in order to predict what policy is necessary for growth depends on accurate data. Falsified data presented by immature narcissistic or sociopathic leaders can release social aggression, the willingness to settle disputes with weapons. That has been Obama's concern.

Wishes for a better life can easily be manipulated. Many people can identify with the false front identity of impostors. In *The Art of the Deal*, written by Tony Schwartz (Random House, 1987), Trump shows how he uses these techniques to win his financial interactions: He has something you want, he scares you about your losing it; and rigs the system against you; he can sue you from safer legal ground. In my psychiatric experience, imposturous people always tell you how they fooled and humiliated you. But if his deals were to flounder, Trump might move toward a totalitarian culture. Thus, as a *New York Times* editorial concluded (November 27, 2016), he would threaten everyone by making an example of Bowe Bergdahl, having him executed for alleged traitorous behavior. The idea of greatness screens an early childhood phantasy of total control over pleasure resources, just as in the early childhood of culture everyone believed their god could become self. When a Pied Piper leads his narcissistic self objects, he will annihilate those who oppose him, as no longer having the right to survive. Just so, the Jihad of Isis represents the self-same cultural syndrome of any authoritarian fundamentalist society. All who joins the leader's Ponzi scheme may imagine complete satisfaction; all in opposition are the irrelevant dead. Such is the story of cultural progress and regress seen as if through Freud's eyes.

References

Aichhorn, A. (1951). *Wayward Youth*, Imago Publishing Co. London.

Akirav, I. & Maroun, M. (2007). The Role of the Medial Prefrontal Cortex-Amygdala Circuits in Stress Effects on the Extinction of Fear. *Neural Plasticity*. 2007, Article ID. 30873.

Atlys, L.Y., et al. (2010). Brain mediators of predictive cue effects on perceived pain, *J Neurosci*. September 29, 2010. 30 (39): 12964–12977. Doi: 10.1523/ JNEURISCI,0057–10.2010.

Beauchaine, T.P., Gatzke-Kopp, L., & Mead, H.K. (2007). Polyvagal theory and Developmental Psychopathology: Emotional Dysregulation and Conduct Problems from Preschool to Adolescence. *Biol Psychol*, February, 2007. 74 (2): 174–184.

Beissner, F., Meissner, K., Bári, K.J., & Napadow, V. (2013). The Autonomic Brain: An Activation Likelihood Estimation Meta-Analysis for Central Processing of Autonomic Function, *J. Neurosci*, 2013 June 19, 33 (25): 10505–10511. Doi: 10:15231/JNEUROSCI.1103–13.2013.

Berlin, H., (2011). The Neural Basis of the Dynamic Unconscious. *Neuropsychoanalysis*, 13 (1).

Birken, L. (1999). Freud's "Economic Hypothesis": From Homo Oecconomicus to Homo Sexualis, *American Imago, 56* (4): 311–330.

Biswal, Bhaat, B. (2010). Toward discovery science of human brain function, ed. Marcus Raichle, *Cross Mark*, 107 (10): 4734–4739. Doi: 10.1073/pnas.0911855107.

Bonnelle, V., et al., & Sharp, D.J. (2012a). ed. Raichle, M., *PNAS*.

Bonnelle, V., et al., & Sharp, V. (2012b). Salience Network Integrity Predicts Default Mode Network Factors after Traumatic Brain Injury, *PNAS*.

Borgogno, F. (2013). *The Girl Who Committed Hari-Kiri: And other Clinical and Historical Essays,* Trans. Alice Spenser, London, Karnac Books.

Boyle, N. (1992). *Goethe: The Poet and the Age*: *Volume I, The Poetry of Desire*. New York, Oxford University Press.

Boyatzis, R.E., Rochford, K., & Jack, A.L. (2014). Antagonistic neural networks underlying differential leadership roles. *Front. Hum. Neurosci.* 8: 114. Doi: 10.3389fnhum.2014.00114. This is an open access article.

Bressler, S.L., & and Menon, V. (2010). Large scale brain networks in cognition: emerging methods and principles, *Trends in Cognitive Sciences*, 14 (6) : 278–290. Doi: 10:1016/j.tics2010.04.004.

Brabant, E., Falzeder, E., & Giampieri- Deutsch, (eds.) (1994). *The Correspondence of Sigmund Freud and Sandor Ferenczi, Volume I: 1908–14*, Trans. Peter H., Cambridge Massachusetts, Harvard Press.

Bullitt, E.D. (1917). *An Uncensored Diary: From the Central Empires*. Garden City, NY, Doubleday, Page and Co.

Bullitt, W.C. (1926). *It's not Done*. New York, Harcourt Brace.

Bullitt, W. (2004). *The Bullitt Mission to Russia,* (Gutenberg, eBook#10713, August 2, 2004)

Bullitt, Ann Moet, Yale University William C. Bullitt Collection [MS 112, Box 30, Folder 680 and Folder 677] Donated by Ann Moet Bullitt, 2008

Burlingham, M. (2002). *Behind Glass: A Biography of Dorothy Tiffany Burlingham*, reprinted from Atheneum, Mcmillan Publishing Group (1989). *The Last Tiffany: A Biography of Dorothy Tiffany Burlingham*, New York, Soft Cover edition, Other Press.

Butler, J. (1990). *Gender Trouble: Feminism and the subversion of Identity*, Routledge, Oxford, U.K.

Carrion, V.G., et al. (2009). Converging evidence for abnormalities of the prefrontal cortex and evaluation of midsagittal structures in pediatric PTSD: an MRI study. *Psychiatry Res.* June 30, 2009 172 (3): 226–234.

Cassela-Blackburn, M. (2004). *The Donkey, The Carrot, and the Club, William C. Bullitt and Soviet-American Relations*, Praeger, Westport Connecticut, U.S.

Cohen, I. (1959). *Theodor Herzl: Founder of Political Zionism.* New York, Thomas Yoseloff.

Cole, M.W., et al., & Braver, T.S. (2012). Global Connectivity of Prefrontal Cortex Predicts Cognitive Control and Intelligence. *The Journal of Neuroscience*, June 27, 2012. 32 (26): 8988–8999. Doi: 10.1523/JNEUROSCI. 10536–12.2012.

Craig, A.D. (2005). Forebrain emotional asymmetry: a neuroanatomical basis? *Trends in Cognitive Sciences*. 19 (12). December 2005.

Craigmyle, N.A. (2013). The beneficial effects of meditation: Contribution of the anterior cingulate and locus Coeruleus, *Front.Psychol.* October 16, 2013. Doi: 10.3389/fps48.2013.00731.

Creswell, R., & Haykel, B. (2015). Battle Lines. *The New Yorker*, (June 8th & 15th 2015): 102–108.

Crittenden, B.M., Mitchell, D.J., Duncan, J. (2015). Recruitment of the default mode network during a demanding act of executive control. Doi: http//dx.doi.org/10.7554/eLife.06481, *eLife* 2015;4:e06481.

Damasio, A. (2003). *Looking for Spinoza*. Harcourt (A Harvard Book) Orlando Florida, U.S.

Dang-Vu, T.T., et al. (2010). Functional Neuroimaging Insights into the Physiology of Human Sleep, (Review), *Sleep*. 33 (12): 1589–1603, 2003.

Danto, E. (2005). *Freud's Free Clinics: psychoanalysis and social justice, 1918–1938*, New York, Columbia University Press.

Darwin, C. (2013). *The Autobiography of Charles Darwin: From The Life and Letters of Charles Darwin* Guttenburg Edition, Ed. by his Son, Francis Darwin (1887).

Darwin, C.*The Descent of Man And Selection in Relation to Sex, 1871*, John Murray Publisher, London. (E Book # 2300, January 6, 2013, Produced by Sue Asscher.)

Davey, C.G., Pujoy, J., Harrison, B.J. (2016). Mapping the self in the brain's default mode network, *NeuroImage*, May 15, 2016, 132: 390–397. Doi:10.1016/j.neuroimage.2016.02.022.

Davies, N. (1996). *Europe: A history*, Oxford University Press, . Harper Perennial Edition, 1998.

Dehaene, S. (2014). *Consciousness and the Brain*: *Deciphering How the Brain Decodes Our Thoughts*, Viking press.

Deen, S. (2015). *All who go do not return,* Greywolf Press.

Diamond, J. (1992). *The Third Chimpanzee*: *the evolution and future of the human animal*, Tokyo (Kodansha-USA-New York) Harper Collins Perennial Edition.

Doi, T. (1971). *The anatomy of dependence*: *The key analysis of Japanese behavior,* Kodancha International Ltd., Tokyo/New York, Transl. John Bester 1973, ninth printing, 1989.

Du, M.-Y., et al. (2015). Altered functional connectivity in the brain default mode network of earthquake survivors persists after two years despite recovery from anxiety symptoms, *SocCognAffectNeurosci.* Doi: 10:1093/scan/nsu040 (2005).

Durkheim, E. (1897) [1951]. *Suicide*. NY, the Free Press.

Eggebrecht, A.T., et al., (2017). Joint Attention and Brain Functional Activity in Infants and Toddlers, *Cereb Cortex*. 1–12. Doi: https://doi.org/10.1093/cercor/bhw403. January 7, 2017.

Eissler, K. (1951). Preface to Aichhorn, A, *Wayward Youth*, London, Imago Publishing Co.

Eldar, E., Cohen, J.D., & Niv, Y. (2013). The effects of neural gain on attention and learning, *Nat.Neuroscie.* August 2013: 16 (8): 1146–1153. Doi: 10.1038/nn3429.

Ellman, S.J., & Antrobus, J.S. (eds.), (1991). *The Mind in Sleep: Psychology and Physiology*, Second Edition. Wiley Interscience Publication.

Ellenberger. (1970). *The Discovery of the Unconscious,* New York, Basic Books.

Fair, D.A., et al. (2008). The maturing architecture of the brain's default network. Cambridge Journals, Medical History, V 16(1) www.pnas.org/cgi/doi/10.1073/pnas.0800501376105.

Farmin, A.S.R. et al., & Yanegishi, T. (2016). Representation of economic preferences in the structure and function of the amygdala and prefrontal cortex, *Scientific Reports*, (*SciRep*. 2016), 6 (20982). Doi: 10.1038/srep20982.

Ferber, M. (2005). *European Romantic Poetry*, Pearson Education Inc., Trans. Jane K. Brown, New York. Pearson Longman, 2005, London.

Ferenczi, S. (1994). *The Correspondence between Sigmund Freud and Sandor Ferenczi. Volume I: 1908–14*, Ed. Brabant, E., Falzeder, E., & Giampieri-Deutsch, P. Trans. Hoffer, P. Harvard University Press, March 1994.

Ferenczi, S. (1933). The Confusion of Tongues between Adults and Children: The Language of Tenderness and of Passion. Sandor Ferenczi Number. M. Balint (Ed.) *International Journal of Psychoanalysis*. 30, Whole No. 4, 1949.

Fornita, A., Harrison, B.J., Zelisky, A. and Simons, J.S. (2012). (ed.) Marcus E. Raichle. Competitive and cooperative dynamics of large-scale brain functional networks supporting recollection, *PNAS*. July 31, 2012, 109 (31): 12788–12793. Doi: 10.1073/pnas.1204185109.

Fox, M.D., et al., & Raichle. (2006). Spontaneous neuronal activity distinguishes human dorsal and ventral attention systems, *PNAS* , Cross Mark. 103 (26): 10046–10051. Doi: 10.1073/pnas.0604187103. May 26, 2006.

Frazer, Sir J.G. (1910). *Totemism and Exogamy* MacMillan and Co. Limited, St. Martin's Street, London (not in copyright).

Freud, S. (1897). *Extracts from the Fliess papers*, S.E. Vol. I.

Freud, S. (1899). *Screen Memories*, S.E. Vol. III. Freud, S. (1950 [1895]). *Project for a Scientific Psychology*, Vol. I.

Freud, S. (1900). *The Interpretation of Dreams*, S.E., Vols. IV & V. Freud, S. (1901). *The Psychopathology of Everyday Life*, S.E.,Vol. VI.

Freud, S. (1909). *Family Romances*, S.E. Vol. IX.

Freud, S. (1909). *Analysis of a Phobia in a Five-Year-Old Boy*, S.E., Vol. X.

Freud, S. (1909). *Notes on a case of Obsessional Neurosis,* S.E., Vol. X.

Freud, S. (1910), *Leonardo Da Vinci and a Memory of his Childhood,*, S.E., Vol. XI.

Freud, S. (1910). *The Antithetical Meaning of Primal Words*, S.E., Vol. XI.

Freud, S. (1911). *The Case of Schreber*, S.E., Vol. XII.

Freud, S. (1914). *The Moses of Michelangelo,* S.E. Vol. XIII.

Freud, S. (1913). *Totem and Taboo*, S.E., Vol. XIII.

Freud, S. (1914). *On the History of the Psychoanalytic Movement*, S.E. Vol. XIV.

Freud, S. (1914). *On Narcissism: An Introduction*, S.E. Vol. XIV.

Freud, S. (1915). *The Unconscious*, S.E., Vol. XIV.

Freud, S. (1916). *Introductory Lectures on Psychoanalysis*, S.E., Vol. XV.

Freud, S. (1917). *A Difficulty in Psycho-Analysis,* S.E., Vol. XVII.

Freud, S. (1917). *A Childhood Recollection from Dichtung Und Wahrheit* (Truth and Fiction), S.E., Vol. XVII.

Freud, S. (1918) [1914]. *From the History of an Infantile Neurosis*, S.E., Vol. XVII.

Freud, S. (1918). *Lines of Advance in Psycho-Analytic Therapy,* S.E., Vol. XVII.

Freud, S. (1919). *The Uncanny*, S.E., Vol. XVII.

Freud, S. (1919). *Victor Tausk* S.E., Vol. XVII.

Freud, S. (1919). *Introduction to Psycho-Analysis and the War Neuroses*: *Appendix: Memorandum of the Electrical Treatment of War Neurotics,* S.E., Vol. XVII.

Freud, S. (1919). *A Child Is Being Beaten,* S.E., Vol. XVII.

Freud, S. (1921). *Beyond the Pleasure Principle*, S.E., Vol. XVIII.

Freud, S. (1922). *Dreams and Telepathy*, S.E., Vol. XVIII.

Freud, S. (1923). *The Ego and the Id*, S.E., Vol. XIX.

Freud, S. (1924). *The Economic Problem of Masochism*. S.E., Vol. XIX.

Freud, S. (1925). Preface to Aichhorn's *Wayward Youth*, S.E., Vol. XIX.

Freud, S. (1926). *Inhibitions Symptoms and Anxiety* S.E., Vol. XX.

Freud, S. (1927). *The Future of an Illusion*, S.E. Vol. XXI.

Freud, S. (1927). *Fetishism*, S.E. Vol. XXI.

Freud, S. (1930) [1929]. *Civilization and its Discontents*, S.E., Vol. XXI.

Freud, S. (1933) [1932]. *New Introductory Lectures* Lecture *on Psychoanalysis* XXXI S.E. Vol. XXII. Freud, S. (1933) [1932]. *New Introductory Lectures on Psychoanalysis, Lecture XXX* S.E., Vol. XXIII.

Freud, S. (1939). *Moses and Monotheism* S.E., Vol. XXIII.

Freud, S., & Bullitt, W.C. (1966). *Thomas Woodrow Wilson: A Psychological Study*, Houghton Mifflin Co.

Gamwell, K., et al., & Jonavovic, T. (2015). Fear conditioned response and PTSD symptoms in children: sex differences in fear-related symptoms, *Developmental Psychobiology*, (Dev/psychobiol.) November, 2015. 57 (7): 799–808.

Gerge, F.A. (2016). *Isis, a History*, Princeton University Press..

Giambrone, J. (2013). Uncle Ruslan Tsarni's Organization may Have Funded Terrorists, *Foreign Policy Journal*, US, Viewpoints, 12, May 3, 2013.

Gilgamesh: The Epic of Gilgamesh 1200 B.C.E., transl. Maurine Gallery Kovaks. Quotes from Gilgamesh are from The Academy of Ancient Texts in the public domain and may be quoted in print without attribution.

Goethe, W. (1848). *Autobiography*: *Truth and Fiction Relating to my Life*. Trans. John Oxenford, London .[Guttenberg Edition.]

Graham, A.M., Fisher, P.A., & Pfeifer, J.H. (2013). What Sleeping Babies Hear: An fMRI Study of Interparental Conflict and Infant's Emotional Processing, *Psychol Sci.* May 1, 2013, 24 (5): 782–789.

Greenberg, R., & Pearlman, C. (1974). Cutting the REM nerve: An approach to the adaptive role of REM sleep. *Perspectives in Biology and Medicine*, 17, 513–521.

Grousset, R. (1970). *The Empire of the Steppes: a History of Central Asia.* Trans. from French by Naomi Walford, Rutgers University Press . Fourth edition, 1994.

Greenacre, P. (1971). *Emotional Growth: Psychoanalytic Studies of the Gifted and a Great Variety of Other Individuals,* Vols. I & II, International Universities Press.

Harris, J.E., & Pontius, A.A. (1975). Dismemberment Murder In Search of the Object, *Psychiatry and Law*, Harris, Jay, and Harris Jean, New York.

Harris, J., & Harris, J. (1981). *The Roots of Artifice: On the Origin and Development of Literary Creativity,* Human Sciences Press.

Harris, J., & Joseph, C. (1973). *Murals of the Mind.* International Universities Press.

Harris, J., & Harris, J. (1984). *The One-eyed Doctor: Sigismund Freud: Psychological origins of his works,* Jason Aronson.

Harris, J.E., & Newman, J. (1986). The Emotional Cost of War (Review Essays), *The Journal of Psychohistory.* 14 (2). Fall, 1986.

Harris, J.E. (1986). *Clinical Neuroscience: From Neuroanatomy to Psychodynamics.* Human Sciences Press.

Harris, J.E. (1998). *How the Brain Talks to Itself: A Clinical Primer of Psychotherapeutic Neuroscience.* The Hawthorn Press.

Harris, J.E., (2013). *Minding the Social Brain*, IP Books, dist. by Jason Aronson.

Hartmann, H. (1964). *Essays on Ego Psychology*. International Universities press.

Hayashi, Y., et al. (2015). Cells of a common origin regulate rem/nonrem sleep and wakefulness in mice, *Science*. October 22, 2015. Doi: 10:1126science.aad1023.

Helm, S. (2016). "Isis in Gaza", *New York Review of Books*, January 14, 2016: 20.

Hulsmann, J.G. (December 29, 2007). *Carl Menger: Pioneer of "Empirical Theory,"* from Chapter 4, *Mises: The Last Knight of Liberalism. Mises Daily*, Mises Institute.

Jacobs, S., Filipov, D., & Wen, P. (2013). *The fall of the House of Tsarnaev, The Boston Globe*, A five months Globe investigation in Russia, Kyrgyzstan, Canada and the United Stated about all the Tsarnaev family members based on multiple based on multiple interviews. December 15, 2013.

Jones, G.A., & Cohen, J.D. (2005). An Integrative Theory of Locus Coeruleus-Norepinephrine Function: Adaptive Gain and Optimal Performance, *Annu. Rev. Neuroscie.* 28: 403–450. Doi: 10.1146/annarev.neuro.28.061604.135709.

Jung, C.G. [1956] (1967). *Symbols of Transformation, Collected Works, Volume 5*, Princeton N.J.: Princeton University Press.

Jung, C.G. (1970). *The Discovery of the Unconscious: The History and Evolution of Dynamic Psychiatry*, New York, Basic Books.

Jung, C. (2012). *The Red Book: Liber Novus*. (ed.) Shamdasani, S. Trans. Kyburz, M., Peck J., & Shamdasani, S. New York, W.W. Norton.

Junger, S. (2016). *Tribe: On Homecoming and Belonging*, New York, Hachette Book Group, Inc.

Kernberg, O. (1975). *Borderline Conditions and Pathological Narcissism*, New York, Jason Aronson.

Koenigs, M., & Grafman, J. (2009). Post-traumatic stress disorder: The role of medial prefrontal cortex and amygdala. *Neuroscientist*, October 15, 2009: 554–548.

Kohut, H. (1979)., The Two Analyses of Mr. Z, *The International Journal of Psycho-Analysis*. 60, Part 1.

Kris, E. (1952). *Psychoanalytic Explorations in Art*, IUP, New York.

Kucyi, A., Hodaie, M., & Davis, K.D. (2012). Lateralization in intrinsic functional connectivity of the temporoparietal junction with salience-and attention-related brain networks, *Journal of Neurophysiology*, December 15, 2012. 108 (12): 3382–3392. Doi: 10.1152/jn.00674.2012.

Leavy, S.A., (1964). Intro and Trans., *The Freud Journal: Lou Andreas-Salome*, Basic Books.

Leknes, S., et al. (2013). The importance of context: When relative relief renders pain pleasant, *Pain*. March 2013. 154 (3): 402–410. Doi: 10.1016/j.pain.2012.11.018.

Lanteaume, L., et al. (June, 2007). "Emotion induction after direct intracerebral stimulations of human amygdala." *Cerebral Cortex*. 17 (6): 1307–1313. Doi: 10.1093/cercor/bhl041. PMID 16880223.

Lasch, C. (1979). *The Culture of Narcissisim: American Life in An Age of Diminishing Expectations*, W.W. Norton and Company, Inc., Warner Books Edition.

Le Doux, J. (2015). *Anxious; Using the Brain to Understand and treat Fear and Anxiety*, Penguin.

Leech, R., Braga, R., & Sharp, D. J. (2012). Echoes of the Brain within the Posterior Cingulate Cortex, *The Journal of Neuroscience* (January 4, 2012): 215–222. Doi: 10.1523/ JNEUROSCI.3689–11.2012.

Levi. P. (1995). *The Reawakening*. Trans. Stuart J. Woolf, Touchstone.

Lipton, E., Confessore, N., & Williams, B. (2016). Top Scholars Or Lobbyists? Often It's Both: Links to Big Business That Go Unseen, *The New York Times*, August 8, 2016.

Liu, H., et al., & Buckner, R. (December 1, 2009) Evidence from intrinsic activity that asymmetry of the human brain is controlled by multiple factors, *PNAS; Crossmark*. 106 (48) Hesheng Liu. 20499–20503. Doi: 10.1073/pnas.0908073106.

Luria, A.R. (1966) [1980]. *Higher Cortical Functions in Man.* Basic Books.

Maddock, R.J., Garrett, A.S., & Buonocore, M.H. (2001). Posterior cingulate cortex Activation by Emotional Words: fMRI Evidence From a Valence Decision Task, *Human Brain Mapping*. 18: 30–41.

Masson, J.M., (1985). *The Complete Letters of Sigmund Freud to Wilhelm Fliess: 1887–1904*, Cambridge MA & London England. The Belknap Press of Harvard University Press.

McGilchrist, I. (2009). *The Master and his Emissary: The Divided Brain and the Making of the Western World.* New Haven and London, Yale University Press, paperback 2010.

McGuire, W., (Ed.) (1974). *The Freud/Jung Letters*, Bollinger Series XCIVI, Princeton NJ, Princeton University press.

Mead, M. (1928). *Coming of Age in Samoa,* William Morrow & Co.

Menon, V. (2015). Large Scale Functional Brain Organization. (ed.) Arthur W.T., *Brain Mapping: An Encyclopedic Reference, vol. 2:* 449–59. Academic press: Elsevier.

Meyer, M.L. et al., & Lieberman, M.D. (2015). Differential neural activation to friends and strangers link interdependence to empathy. *Cult. Brain.* Doi: 10.1007/ s40167–014–0023–7.

Molochnikov, I., & Cohen, D. (2014). Hemispheric differences in the mesostriatal dopaminergic system, *Frontiers in Systems Neuroscience.* 8: 110. ncbi.nim.nih.gov.

Moran, J.M., Kelley, W.M. & Heatherton, T.F. (2013). What Can the Organization of the Brain's Default Mode Tell Us About Self-Knowledge? *Frontiers in Human Neuroscience*, July, 2013. 7: 347. Doi:10.3389/fnhun.2013.00391.

Nielson, J. A., et al. (2013). An Evaluation of the Left-Brain vs. Right-brain Hypotheses with Resting State Functional Connectivity Magnetic Resonance Imaging. Doi: 10.1371/journalpone.007.1275.

Northoff, G., et al. (2011). The resting state hypothesis of Major Depressive Disorder—a Transactional Subcortical-cortical Framework for a System Disorder, *Neuroscience and Behavioral Reviews.* (35) 9: 1929. Doi: 10.1016/ neurobiorev.2010.12.0.

O'Donnell, M.B., Falk, E.B., & Lieberman, M.D. (2015). Social in, social out: How the brain responds to social language with more social language, *Communication Monographs,* Routledge. Doi: 10.1080/03637751.2014.990472.

Panksepp, J. (2010). Affective neuroscience of the emotional BrainMind: Evolutionary perspectives and implications for understanding depression. *Dialogues Clin Neurosci.* December 20, 2010. 12 (4): 533–545.

Panksepp, J. (2011). Cross-Species Affective Neuroscience Coding of the Primal Affective Experiences of Humans and Related Animals. *PLoS One.* 6 (9): e2136. Doi: 10:1371/journal.pone.002136.

Phillips, A. (1994). *Secrets, The London Review of Books.* 16 (19), October 6, 1994.

Porges, S.W. (1993). "Orienting in a defensive world: Mammalian modifications of our evolutionary heritage. A Polyvagal Theory", *Psychophysiology.* 32: 301–318.

Porges, S.W. (2011). *The Polyvagal Theory: Neurophysiological Foundations of Emotions, Attachment, Communication and Self-Regulation.*W.W. Norton & Company..

Raichle, M.E. (2010). Two Views of Brain Function. *Trends in Cognitive Sciences.* 14 (4): 180–190. Doi: 10.1016/jtics.2010.01.008.

Raichle, M. (2015). The restless brain: how intrinsic activity organizes brain function, *Philosophical Transactions of the Royal Society*, B370: 20140172. http://dr.- 0172 Published March 30, 2015 (Creative commons attribution License) doi.org/10.1098/rstb.2014.

Rakoff, J.S. (2016). "The Magna Carta Betrayed?" *The New York Review of Books*, February 11, 2016.

Rakoff, J.S. (2016). "Neuroscience and the Law: Don't Rush In" *The New York Review of Books,* May 12, 2016.

Rank, O. (1904). *The Myth of the Birth of the Hero: and other writings.* Vintage books. (See S.E. Vol. XXIII, Bibliography, p. 310.)

Reitman, J. (2013). Jahar's World, *Rolling Stone Magazine*, July 17, 2013.

Reynolds, M.A. (2013). *The Northern Caucasus, the Tsarnaevs, and Us* (May 6, 2013), From the Foreign Policy Research Institute, Archives, E-Notes, fpri.org.

Richert, K.A., Carrion, V.G., Karchemskiy, A., and Reiss, A.L. (2006). Regional differences of the prefrontal cortex in pediatric PTSD: an MRI study, *Depress Anxiety*. 23 (1): 17–25.

Roazen, P. (1975). *Freud and His Followers*, New York, Knopf.

Roosevelt, F.D. (March 4, 1933), Franklin Roosevelt's First Inaugural Address.

Ruthven, M. (2016). *How to Understand Isis*, "Review of The New Arab Wars: Uprisings and Anarchy in the Middle East" (ed.) Lynch M., *Public Affairs* and "Isis: A History" (ed.) Gerges, F.A. *Princeton University Press New York Review of Books*, June 23, 2016, LXIII (11): 20–24.

Salome, L.A. (1964). *The Freud Journal of Lou Andreas-Salomé*, Trans. & Intro. Leavy, S.A. New York, Basic Books, Inc.

Samuel, L.R. (2013). *Shrink: A Cultural History of Psychoanalysis in America,* University of Nebraska Press.

Santner, E.L. (1996). *Daniel Paul Schreber's Secret History of Modernity,* Princeton, NJ, Princeton University Press, 08540.

Sartre, J.P. (1963). *The Word.*, Trans. Bernard Frechtman & George Brazillar.

Sartre, J.P. (1988). *Saint Genet,* Pub: William Heinemann Ltd.

Schorske, C. (1980). *Fin-De-Siécle Vienna: Politics and Culture,* Alfred A. Knopf, New York.

Seelye, K.Q., & Bidgood, J. (2015). Breaking Silence, Dzhokhar Tsarnaev Apologizes for Boston Marathon Bombing, *The New York Times*, June 24, 2015.

Shamay-Tsoory, S.G., et al. (2009). Intranasal administration of oxytocin, November 1, *Biol Psychiatry*. 66 (9): 864–70.

Simony, E., et al. (2016). Dynamic reconfiguration of the default mode network during narrative comprehension, *Nature Communications* 7, July 18, 2016, published online. Doi: 10.1038/ncomms12141.

Sledge, W. (1992). *Journal of Psychotherapy Practice and Research*. 1 (2), Spring, 1992 (ncbi.nim.nih.gov).

Smith, W.R. (1927). *Lectures on the Religion of Semites*, 1894. The Macmillan Co. contributed by the Universal Digital Library.

Solms, M., & Panksepp, J. (2012). The "Id" Knows More than the "Ego" Admits: Neuropsychoanalytic and Primal Consciousness Perspectives on the Interface Between Affective and Cognitive Neuroscience. *Brain Sci*. 2012. 2: 147–175. Doi: 10.3390/rainsci2020147.

Spoormaker, V.I., et al. (2014). Ventromedial prefrontal cortex activity and rapid eye movement sleep are associated with subsequent fear expression in human subjects. *Experimental Brain Research*, May 2014. 232 (5): 1547–1554.

Spunt, R.P., Meyer, M.L., & Lieberman, M.D. (2015). The Default Mode of Human Brain Function Primes the Intentional Stance, *Journal of Cognitive Neuroscience*. 27 (6): 1116–1124. Doi: 10.1162/jocn_a_00785.

Sridharan, D., Levitin, D.J., and Menon, V.A. (2008). A Critical Role for the right fronto-insular-cortex in switching between central executive and default mode networks, *PNAS* USA, August 26, 2008: 105(34): 12589–94. Epub August 22, 2008.

Stevens, W. (1956). *The Collected Poems of Wallace Stevens*, Alfred A. Knopf, New York.

Stewart, W.A., & Freeman, L. (1972). *The Secret of Dreams*, Macmillan Publishing Co.

Streissler, E.W. (1990). "Carl Menger on Economic Policy: The Lectures to Crown Prince Rudolf": *History of Political Economy*, 22 (annual supplement): 107–112.

Strozier, C. (2001). *Heinz Kohut: The Making of a Psychoanalyst*, Farrar, Straus and Giroux, New York.

Swanson, N., Eichele, T., & Calhoun, V.D. (2011). Lateral Differences in the Default Mode Network in Healthy Controls and Schizophrenic Patients, *Hum Brain Mapp*; April. 32 (4): 654–664.

Tacikowski, B., & Ehrsson. (2017). Dissociating the Neural Base of Concetual Self-Awareness from perceptual Awareness and unaware Self-Processing, Cerebral Cortex, 1–14. Doi: 10.1093/cercor/bhx004.

Tausk, V. (1990). *Sexuality, War and Schizophrenia: Collected Psychoanalytic Works* London,Transaction Publisher, Karnac Books.

Tausk, V. (1919). Origin of the Influencing Machine in Schizophrenia, *Psychoanalytic Quarterly*, 1933, English Trans., Dorian Feigenbaum.

Van den Heuvel, M.P. & Sporns, O. (2013). Network hubs in the human brain, *Trends in Cognitive Sciences*. 17 (12): 683–696, December, 2013. Doi: http//dx.doi:org/10.1016/jtics.2013.09.012.

Van den Stock, et al. (2015). *Sci. Rep*. 5, 13779. Doi: 10.1038/srep13779.

Utevsky, A.V., Smith, D.V. & Heuttel, S.A. (2014). Precuneus Is a Functional Core of the Default Mode Network, *J. Neurosci.* January. 15. 34 (3): 932–940. Doi: 10.1523/JNEUROSCI.2247–13.2014.

Watson, K.K., Jones, T.K. and Allman, J.M. (2006). Dendritic Architecture of the Von Economo Neurons. *Neuroscience.* 141 (2006): 107–112.

Welborn, B.L. & Lieberman, M.D. (2015) . Person-Specific Theory of Mind in medial prefrontal cortex, *Journal of Cognitive Neuroscience,* 27: 1–12.

Williams, W.C. (1955). *Journey to Love.* Random House, New York.

Winnicott, D.W. (1965). "Ego distortion in terms of true and false self", *The maturational Process and the Facilitating Environment*, International University Press.

Wundt, W. (1916). *Elements of Folk Psychology.* Project Gutenberg E books prepared by Clare Graham, Heather Strickland, and Mark D'Houghe. Trans. Edward Leroy Schaub, Ph.D.

Yagi, K. (2011). *Austrian and German Economic Thought: From subjectivism to social evolution*, Routledge Studies in the History of Economics, NY.

Young-Bruehl, E. (1988). *Anna Freud: A Biography*, Yale University press, New Haven and London. (Preface to second edition and revision, 2008.)

Index

About the Author

After his early undergraduate days in psychology and writing at Johns Hopkins University, Jay Harris was intent on becoming a psychiatrist. Working his way through med school, with NIMH summer fellowships, he also had a Saturday job doing mental status exams on new state hospital patients who manifested mental syndromes—before the era of major tranquilizers. His residency, inpatient fellowship, and psychoanalytic training provided exposure to the expertise and hubris of the elite culture of psychoanalysis at the time. That contrasted with his late 1970s and 1980s work in the Bellevue Prison ward and the NY City Department of Welfare's CSS (Community Support Services) project to rehabilitate the homeless, and a decade as a ward chief at New York's Metropolitan Hospital. Writing about the cultural mandates of compassion that yielded to self-centered leadership, he studied neuropsychoanalysis as a humanitarian discipline. In the 1990s, working as Residency Director at Stony Brook University Medical Center and consulting psychiatrist at the University Counseling Center, he helped residents understand why they wanted to become psychiatrists; and counseled doctoral students in connecting their personal identity-organizing experiences to their thesis work. *How Culture Runs the Brain,* Harris's eighth book, takes up Freud's invitation to put cultural syndromes into a scientific frame.